Material Science and Metallurgy

Material Science and Metallurgy

Prof. A.V.K. Suryanarayana
Formerly Head, Metallurgical Section,
V.I.S.S.J. Polytechnic,
Bhadravathi.

BSP **BS Publications**
A unit of **BSP Books Pvt., Ltd.**

4-4-309, Giriraj Lane, Sultan Bazar,
Hyderabad - 500 095 - TG
Phone : 040 - 23445605, 23445688

Published by :

BSP **BS Publications**

A unit of **BSP Books Pvt., Ltd.**

4-4-309, Giriraj Lane, Sultan Bazar,
Hyderabad - 500 095 - TG INDIA
Phone : 040 - 23445605, 23445688
e-mail : info@bspbooks.net
www.bspublications.net

ISBN: 978-93-85433-47-4 (HB)

Dedicated to
my wife
Smt. Annapurna

Preface

Engineers use different materials in service. Which material is suitable to what purpose is the most important aspect of the engineer's analysis. In fact the engineer is for that purpose only. Thus, the subject occupies a very important part of the syllabus of a mechanical or production engineer.

The portion of metallurgy is dealt with utmost care to make the fundamentals clear. Many heat treatment processes are given with an explanation of why it is so done.

Various materials in use are dealt with in an orderly fashion.

Many books are available with their title written to cater to the need of the students of some universities and autonomous institutions. An effort is made herein to make the book suitable to a majority of the syllabi of different universities. Composite materials is written touching different materials with the help of Dr. N. Naik of the department of Aerospace Engineering of the Indian Institute of Science, Bangalore. I am highly indebted to him.

All said and done, the owners of M/s. B.S. Publications, Hyderabad, should be thanked for waiting for years for the script of this book.

The reader can study separately, if a particular necessity arises for elaboration.

All suggestions are welcome.

- Author

Contents

CHAPTER 1

Atomic Arrangement

All materials consist of basically atoms. These atoms are arranged in a regular fashion. This arrangement is called a Space Lattice (Fig. 1.1).

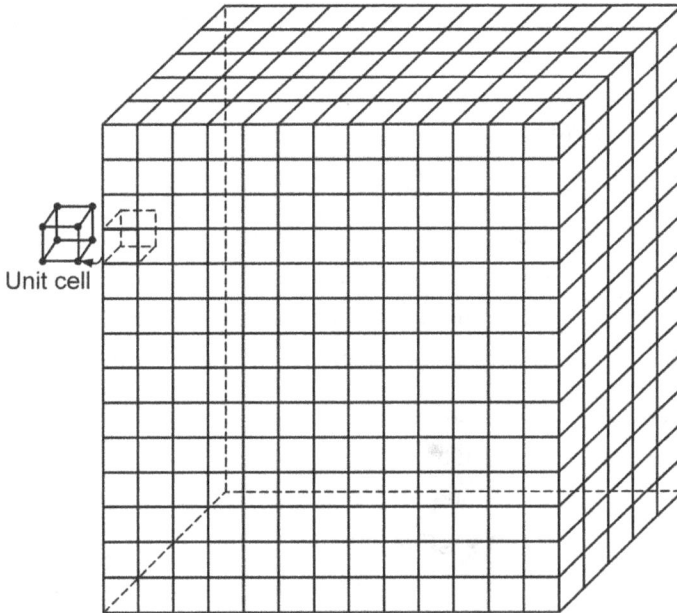

Fig. 1.1 Space lattice.

In this arrangement atoms occupy sites, which are regular and predictable. In other words, the distance between two atoms is equal in all directions. The minimum number of atoms taken as a shape (e.g., a cube), which repeats itself in all the three directions in the lattice is called a 'unit cell'. The shape of the unit cell defines the atomic structure. If the unit cell is a cube, the structure of the lattice is defined as cubic and if it is rhombic, the structure is referred to as rhombic and so on.

The atoms do not occur in a periodic fashion over long distances. Their direction changes after some distance. This changed direction continues for a certain distance and again the direction alters. The existence in a particular direction is called orientation. The region of uniformity is called a crystal and the altered direction in the adjacent region is another crystal. Oftentimes the changes in the orientations are not abrupt but gradual (over atomic distances). These regions between two crystals are called grain boundaries. This crystalline nature of metals is not visible to naked eye. The metal surface should be polished to a mirror finish and suitably etched with very weak acid solution and observed under a microscope at a magnification of at least 100. The study of metals and alloys under microscope is called metallography. The regions between two grains, are revealed as lines separating the grains and these are called grain boundaries.

When we observe two specimens under the microscope at the same magnification and when the grains in one appear bigger than the other, we say that the former metal has a coarse grained structure. The structure with more number of grains under view is called *fine grained*. Reporting like this is arbitrary. There are prescribed methods to define the grain size.

As seen above, the lattice orientation in adjoining grains will be different. It is also noted that the change in orientation is gradual. Thus, the grain boundary region will not contain atoms in any stipulated (definable) pattern. These grain boundaries are regions of disturbance. Further, any impurities in the metal will concentrate at the grain boundaries adding to the atomic disturbance.

Grain Structure

So far, we studied about the space lattice taking the example of a cubic cell. The 'unit cell' need not always be cubic. It is defined by its parameters — its edges a, b, and c and angles α (between b and c), β (between a and c) and γ (between a and b). Fig. 1.2 serves to visualize this better. There are fourteen types of space lattices and they fall into seven crystals systems listed in   Table 2.1.

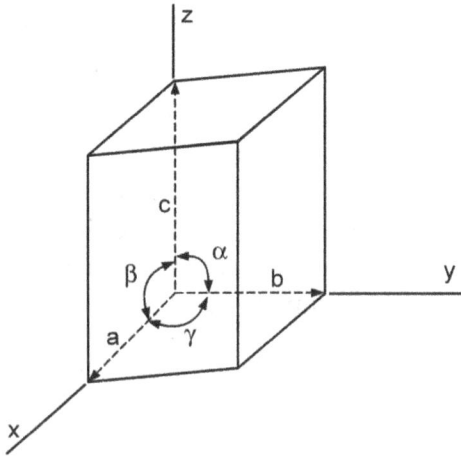

Fig. 1.2 Space lattice illustrating lattice parameters.

Table 1.1 Crystal systems.

S.No.	Lattice	Description
1.	Triclinic	Three unequal axes, no two of which are perpendicular $a \neq b \neq c$; $\alpha \neq \beta \neq \gamma \neq 90°$
2.	Monoclinic	Three unequal axes, one of which is perpendicular to the other two $a \neq b \neq c$; $\alpha = \gamma = 90° \neq \beta$
3.	Orthorhombic	Three unequal axes, all perpendicular $a \neq b \neq c$; $\alpha = \beta = \gamma = 90°$
4.	Rhombohedral (trigonal)	Three equal axes, not at right angles $a = b = c$; $\alpha = \beta = \gamma \neq 90°$
5.	Hexagonal	Three equal coplanar axes at 120° and a fourth unequal axis perpendicular to their plane $a = b \neq c$; $\alpha = \beta = 90°$;$\gamma = 120°$
6.	Tetragonal	Three perpendicular axes, only two equal $a = b \neq c$; $\alpha = \beta = \gamma = 90°$
7.	Cubic	Three equal axes, mutually perpendicular $a = b = c$; $\alpha = \beta = \gamma = 90°$

*From C. S. Barrett, "Structure of Metals," McGraw-Hill Book Company, Inc., New York, 1952.

Again it is our fortune that almost all the metals crystallize into three important crystal structures. These are Face Cetered Cubic (FCC), Body

Centered Cubic (BCC) and Hexagonal Close Packed (HCP). These are illustrated in Fig. 1.3, 1.4 and 1.5.

Face Centered Cubic (FCC): This structure consists of a cube with eight atoms situated at the eight corners (Fig. 1.3). It also consists of an Atom at the centers of each of the six faces of the cube (and hence the name). In this structure, each of the corner atoms shared by eight cubes. Each of the face centered atoms shared by two cubes. So we can calculate the atomic density of the structure thus,

In one lattice, 8 corner atoms $8 \times \dfrac{1}{8} = 1$

In one laatice 6 face center atoms $6 \times \dfrac{1}{2} = 3$

$$\text{Total} = 4 \text{ atoms}$$

The unit cell of FCC structure consists of 4 atoms.

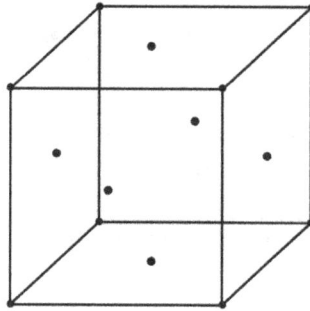

Fig. 1.3 F.C.C. lattice.

Body Centered Cubic: This structure consists of a cube with eight atoms situated at its eight corners (Fig. 1.4). There also exists an atom at the center of the cube. Thus,

In one lattice, 8 corner atoms $8 \times \dfrac{1}{8} = 1$

In one lattice, 1 central atom $= 1$

$$\text{Total}   = 2 \text{ atoms}$$

The unit cell of BCC structure contains 2 atoms.

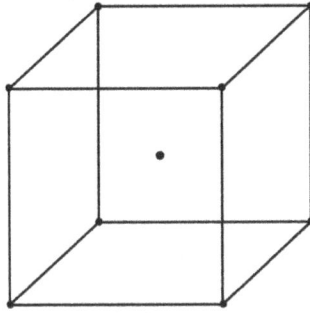

Fig. 1.4 B.C.C lattice.

It can be seen from the above that the FCC cell is more densely packed than a BCC cell. Metals like Chromium, Tungsten, Molybdenum, Vanadium, Sodium and α and δ irons possess this structure. Metals like Aluminium, Copper, Gold, Silver, lead and γ iron are some examples possessing FCC structure.

Hexagonal Close Packed: A hexagonal close packed lattice shows two basal planes which are regular hexagons containing an atom at each of the hexagon corners and one at their geometrical centers (Fig. 1.5). In addition to these atoms, each hexagon contains three atoms at the corners of a triangle in the center of prism. Their location will be inside the hexagon and inside the alternate rectangular sides.

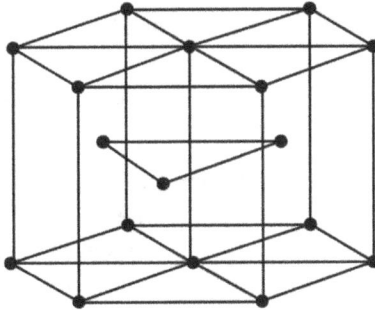

Fig. 1.5 H.C.P lattice.

Since, each corner atom of the unit cell is shared by eight cells, the basal atoms shared by two and the three atoms inside the hexagon are exclusively of the lattice, the atomic density of the lattice is computed as.

12 corner atoms	$12 \times \dfrac{1}{6} = 2$
2 basal atoms	$2 \times \dfrac{1}{2} = 1$
3 central atoms	$= 3$
Total	$= 6$

Examples of metals that crystallize in this type of structure are magnesium, beryllium, zinc, cadmium etc.

The unit cell of a cubic system can be specified by a single lattice parameter i.e., the side of the cube, A hexagonal cell requires the width of the hexagon a and the height (the distance between the two basal planes), c. The axial ratio $\dfrac{c}{a}$ which is sometimes given. The axial ratio for beryllium is given as 1 .58 and for cadmium 1.88.

POLYMORPHISM AND ALLOTROPY

Polymorphism is the property of the material to exist in more than one type of crystal lattice in solid state. If the change (of structure) is reversible, the polymorphic change is called allotropy. About a dozen metals are known to show this allotropic changes, but the most important and best known is iron. Iron solidifies at 1532 °C as δ – iron BCC, while cooling further, the δ – iron changes to γ – iron at 1400 °C which again becomes α – iron at 910 °C. The γ iron has an FCC structure while the α – iron has again BCC structure.

Metal (or alloy) behavior under stress is dependent upon the effect the stress produces on it. This is the sum of the effect on the individual grains of the metal. So, to have a fundamental concept of the metal deformation, we need to understand the deformation of the single crystal. The slip undergone by the single crystal depends upon the magnitude of the shearing stress produced by the load applied, the geometry of the crystal structure and the orientation of the active slip planes with respect to the shearing stresses. Slip starts when the shearing stress on the slip plane in the slip direction reaches an optimum value called the critical resolved shear stress. The value is really the single crystal equivalent of the yield stress in an ordinary stress-stain curve. The value of the critical resolved shear depends upon the composition and temperature.

Schmid postulated that different tensile loads are necessary to produce slip in single crystal of different orientations can be rationalized by critical resolved shear stress. It is necessary to know from X–ray diffraction, the orientation with respect to the tensile axis of the plane on which slip first occurs and the slip direction of the single crystal tested in tension. Let us consider a cylindrical single crystal of cross sectional area A (Fig. 1.6). The angle between the normal to the slip plane and the tensile axis is 'φ' and the angle which the slip direction makes with the tensile axis is 'λ'. The area of the slip plane inclined at an angle 'φ' will be A/cosφ and the component of the axial load acting on the slip plane in the slip direction is P cosλ. Therefore, the critical resolved shear stress is given by

$$\tau r = \frac{p\cos\lambda}{\dfrac{A}{\cos\phi}} = \frac{P}{A}\cos\varphi\cos\lambda$$

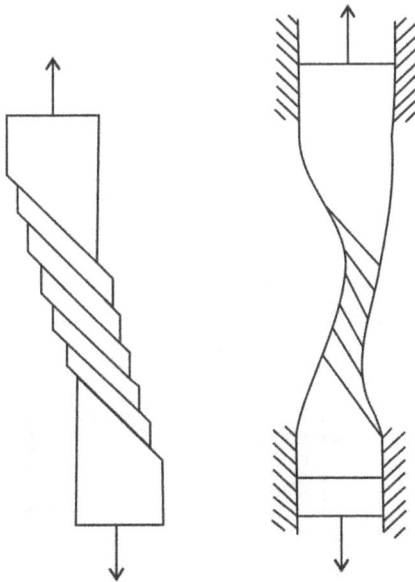

Fig. 1.6

The above equation gives the shear stress resolved on the slip plane in the slip direction. This shear stress is a maximum when φ = λ = 45°, so that $= \frac{1}{2}\left(\frac{P}{A}\right)$. If the tension axis is normal to the slip plane $\lambda = 90°$ or if it is parallel to the slip plane the resolved shear stress is zero. Slip will not recur for these extreme orientations since there is no shear stress on the slip plane. Crystal close to these orientations tend to fracture rather than slip.

It was experimentally observed that small amounts of impurities in the metal increase the critical resolved shear stress. Addition of alloying elements was found to be of even greater effect. Still greater would be the rise in the value of the critical resolved shear stress in an alloy wherein the solute atoms differ considerably from the solvent atoms in size.

The magnitude of the critical resolved shear stress of a crystal is determined by the interaction of the population of dislocations with each other and with defects such as vacancies, interstitials and impurity atoms. It should be understood that when a positive dislocation moves and encounters a negative dislocations, they cancel out. This stress is of course greater than the stress needed to move a single dislocation yet much lower than the stress required to produce slip in a perfect lattice. On the basis, the critical resolved shear stress should decrease as the density of defects decreases, provided that the total number of dislocations is not zero. When the last dislocation is also eliminated, the critical resolved shear stress should rise abruptly. This rise will be equal to the high value predicted for the shear strength of a perfect crystal.

Most studies of the mechanical properties are made, by subjecting the crystal to simple uniaxial tension. In the test, the movement of the cross head of the testing machine constrains the specimen at the grip since the grips must remain in line. As a result, the specimen is not permitted to deform freely by uniform glide on every slip plane along its gauge length. Instead, the slip planes rotate towards the tensile axis since the tensile axis of the specimen remains fixed as in (b). These experimental errors can be compensated by tedious calculations, and it is finally concluded that these can be ignored because of the multitude of grains occurring in the metal specimen under test. This basic concept of deformation at the single crystal level goes a long way in understanding the breakage of the specimen.

CRYSTAL DEFECTS

It is possible to calculate the theoretical strength of a metal by the force required to separate the bond between adjacent atoms. This turns out to be several millions Kgf/cm². Ordinarily the strength of metals is 100 to 1000 times less. The reason for this lies in the occurrence of defects in the crystal structures of metals.

Real crystal do not contain the atoms in perfect symmetry. Some regions will have defects or imperfections. When the deviation from the periodic arrangement of the atoms of the lattice is localized to the vicinity of a few atoms, the defect is called point defect or point imperfection. The lattice imperfections can again be divided into line defects and surface or plane defects.

Point Defects: These are mainly of three types – the vacancy, the substitutional atoms and the interstitial atom. When an atom is found missing from a normal lattice position, a vacancy exists. A number of vacancies can be caused in a pure metal by thermal excitation and these are thermodynamically stable at all temperatures above absolute zero.

$$\frac{n}{N} = e\frac{-Es}{KT}$$

where,

n = No. of vacancies at the absolute temperature T

N = Total no. of lattice sites of atoms.

Es = Energy required to move an atom from interior to the surface

K = Boltzman's constant

An atom that is trapped inside a crystal lattice at a point intermediate between normal lattice positions is called an Interstitial atom (Fig. 1.7). The presence of an impurity atom at a lattice position is when there occurs vacancy or the energy requirements are fulfilled. Bother interstitial and impure atoms result in disturbance of the periodicity of the lattice.

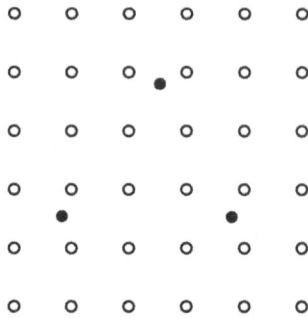

Fig. 1.7 Interstitial atom.

LineDefects/Dislocations: Dislocation is the most important two-dimensional or line defect. This can be imagined to be the region of disturbance between two, otherwise perfect lattice structures. This can also be said as the boundary between the slipped and unslipped regions of a crystal. The discontinuation of a row of atoms (in section) is called and edge dislocation (Fig. 1.8). The slip direction makes with the tensile axis is 'λ'.

The area of the slip plane inclined at an angle 'φ' will be A/cosφ and the component of the axial load acting on the slip plane in the slip direction is P cosλ. Therefore, the critical resolved shear stress is given by

$$\tau r = \frac{P\cos\lambda}{\dfrac{A}{\cos\phi}} = \frac{P}{A}\cos\varphi\cos\lambda$$

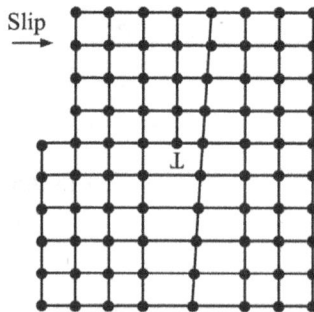

Fig. 1.8 Edge dislocation.

The above equation gives the shear stress resolved on the slip plane in the slip direction. This shear stress is a maximum when $\varphi = \lambda - 45°$, so that $\tau r = \dfrac{1}{2}\left(\dfrac{P}{A}\right)$. If the tension axis is normal to the slip plane $\lambda = 90°$ or if it is parallel to the slip plane ($\varphi = 90°$), the resolved shear stress is zero. Slip will not recur for these extreme orientations since there is no shear stress on the slip plane. Crystals close to these orientations tend to fracture rather than slip.

It was experimentally observed that small amounts of impurities in the metal increase the critical resolved shear stress. Addition of alloying elements was found to be of even greater effect. Still greater would be the rise in the value of the critical resolved shear stress in an alloy wherein the solute atoms differ considerably from the solvent atoms in size.

The magnitude of the critical resolved shear stress of a crystal is determined by the interaction of the population of dislocations with each other and with defects such as vacancies, interstitials and impurity atoms. It should be understood that when a positive dislocation moves and encounters a negative dislocations, they cancel out. This stress is of course greater than the stress needed to move a single dislocation yet much lower than the stress required to produce slip in a perfect lattice. On this basis, the critical resolved shear stress should rise abruptly. This rise will be equal to the high value predicted for the shear strength of a perfect crystal.

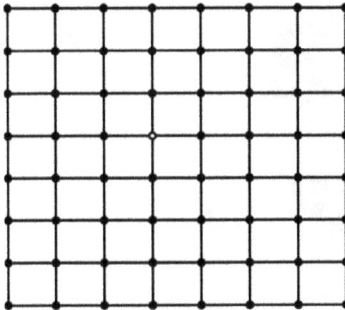

Fig. 1.9 Vacancy.

Most studies of the mechanical properties are made, by subjecting the crystal to simple uniaxial tension. In the test, the movement of the cross head of the testing machine constrains the specimen at the grip since the grips must remain in line. As a result, the specimen is not permitted to deform freely by uniform glide on every slip plane along its gauge length (Fig. 1.10). Instead, the slip planes rotate towards the tensile axis since the tensile axis of the speciemen remains fixed.

These experimental errors can be compensated by tedious calculations, and it is finally concluded that these can be ignored because of the multitude of grains occurring in the metal specimen under test. This basic concept of deformation at the single crystal level goes a long way in understanding the breakage of the specimen.

GRAIN SIZE

Many times the words coarse grain size, fine grain size and grain boundaries come up in discussions regarding metal structures. A fine grained structure as it appears under a microscope at a magnification of say 100 X will appear coarse when viewed at a magnification of 200 or 500 X. Signifying a grain size in a way that a comparison can be made with others or the same or similar grain is replicated is a real task. A rational method is required to define a grain size. American society for testing materials had evolved a method of determining the grain size of metals or alloys, by classifying different grain patterns by their exact size by giving them numbers. The grain size, so calculated is known as the ASTM grain size. [Fig. 1.10 (a) to (b)]

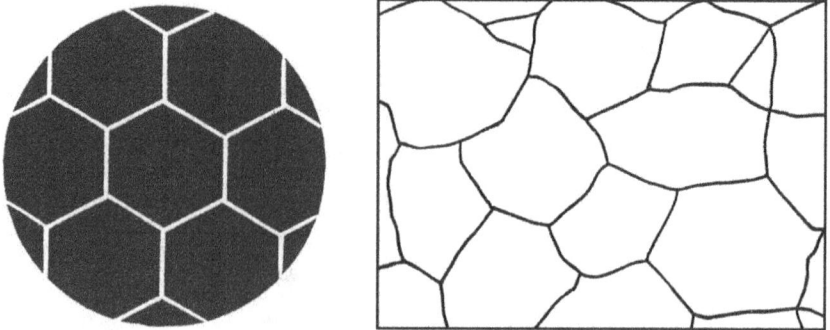

Fig. 1.10(a) Upper, idealized hexagonal network for mean grain size No. 1, ASTM scale, 1 gr per sq in. Lower, ASTM standard grain size No. 1. Up to $1^{1/2}$ gr per sq in. at 100 X.

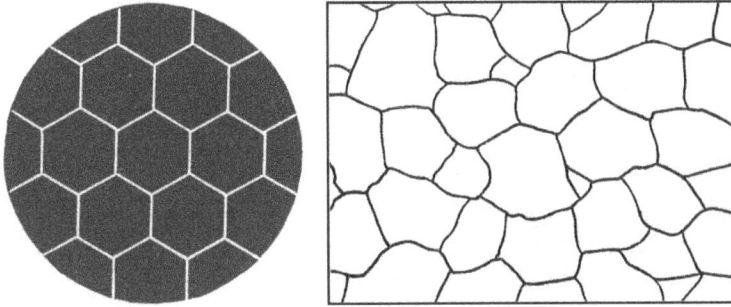

Fig. 1.10(b) Upper, idealized hexagonal network for mean grain size No. 2,
ASTM scale, 2 gr per sq in. Lower, ASTM standard grain size No. 2, $1^{1/2}$ to 3 gr
per sq in. at 100 X.

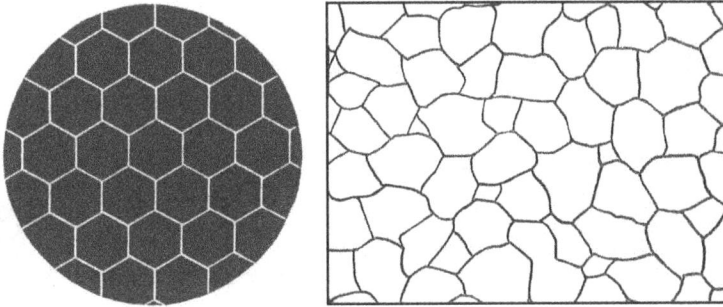

Fig. 1.10(c) Upper, idealized hexagonal network for mean grain size No. 3,
ASTM scale, 4 gr per sq in. Lower, ASTM standard grain size No. 3, 3 to 6 gr
per sq in. at 100 X.

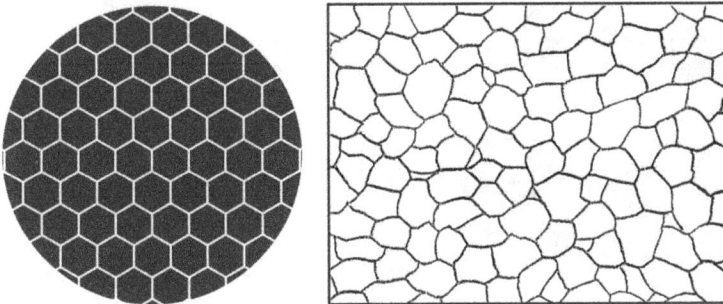

Fig. 1.10(d) Upper, idealized hexagonal network for mean grain size No. 4,
ASTM scale, 8 gr per sq in. Lower, ASTM standard grain size No. 4, 6 to 12 gr
per sq in. at 100 X.

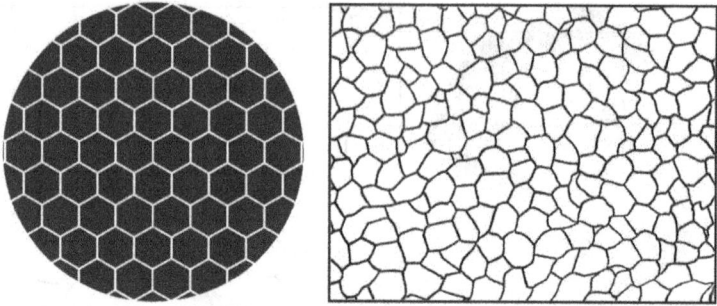

Fig. 1.10(e) Upper, idealized hexagonal network for mean grain size No. 5, ASTM scale, 16 gr per sq in. Lower, ASTM standard grain size No. 5, 12 to 24 gr per sq in. at 100 X.

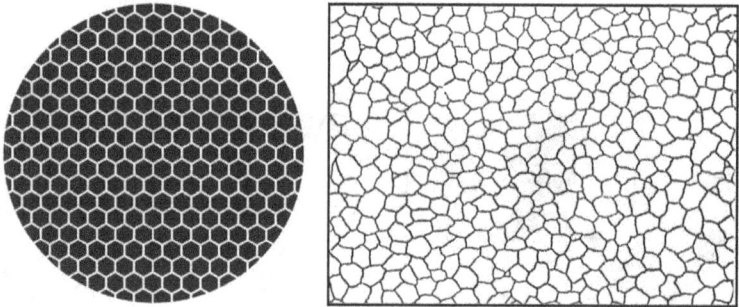

Fig. 1.10(f) Upper, idealized hexagonal network for mean grain size No. 6, ASTM scale, 32 gr per sq in. Lower, ASTM standard grain size No. 6, 24 to 48 gr per sq in. at 100 X.

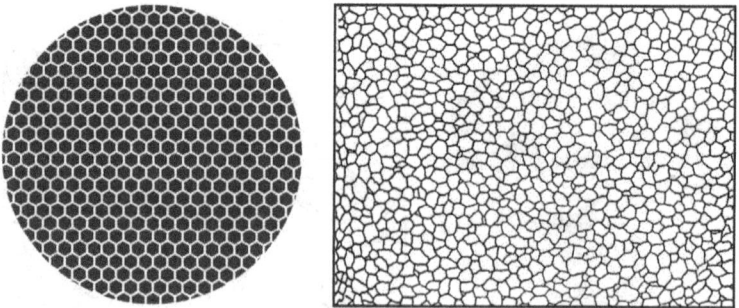

Fig. 1.10(g) Upper, idealized hexagonal network for mean grain size No. 7, ASTM scale, 64 gr per sq in. Lower, ASTM standard grain size No. 7, 48 to 96 gr per sq in. at 100 X.

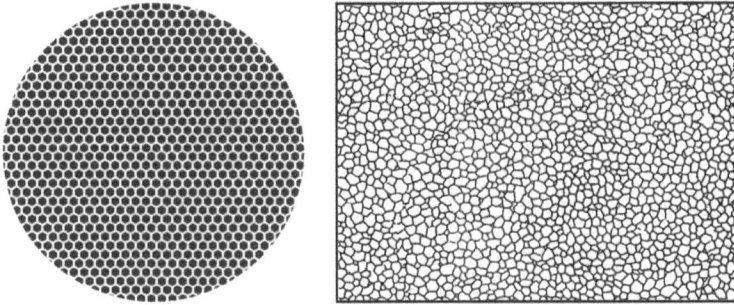

Fig. 1.10(h) Upper, idealized hexagonal network for mean grain size No. 8, ASTM scale, 128 gr per sq in. Lower, ASTM standard grain size No. 8, 96 to 192 gr per sq in. at 100 X.

Definition

The grain size of a metal or alloy is defined as the number of grains of the metal or alloy in an area of one square inch, viewed at a magnification of 100 X. An empirical relationship is as follows,

$$N = 2^{n-1}$$

where n = number of grains of the metal (or alloy) at a magnification of 100 X

and   N = the grain size number.

Many metals and alloys can be polished, etched with a suitable reagent and viewed under a microscope. But some microstructures are difficult to view with a clear grain pattern so that their number can be counted. Many steels are austenitised and held at a sub critical temperature for a very short (pre-determined) time so that pearlite, cementite or bainite starts to form at the austenitic grain boundaries and then quenched. It is known that decomposition of austenite starts at the grain boundaries. Later, the quenched specimen can be polished, etched and viewed under the microscope at 100 X and the number of grains in an area of one square inch computed.

To avoid the cumbersome procedure of counting the number of grains in a square inch area, A.S.T.M had standardized 8 grain sizes and gave them in the form of numbered figures. These standard charts give ideally sketched grains pattern and a metallographic structure of the same grain size 1.1 to 1.8 for really use. This is very handy method.

Fracture Method

This method improves the macro comparison. Standard fractures of steels are developed and they are numbered from 1 to 10. Shepherd and Jern Konteret are the scientists who first developed this comparison of fractures method for alloy steels.

The process involves the use of a standard 1" dia test bar which is cut (slotted) to a small depth of $\frac{1}{16}$" transversely. The bar is then hardened. It is held in a vice and a hammer blow is imparted to the protruding portion of the bar in order to break it at the slot.

The fracture surface reveals granular structure. The surface is compared with the standard fractures to match the size of the grains. The number of the matching standard fracture is reported as the grain size of the steel by this method.

The grain sizes reported by the macro (fracture) method or by the microscopic method will tally to a large extent. Steels with grain sizes less than four are generally called coarse and the other four are called medium and fine. The grain size with number 8 is also rare.

Manifestation/ Property	Fine grained structure	Course grained structure
1. Depth of hardening	Less	More
2. Retained austenite (on quenching)	Less	More
3. Warpage (on quenching)	Less	More
4. Quench cracks	Less	More
5. Internal stresses	Less	More
6. Machine ability after Normalizing	Inferior	Better
7. Toughness	More	Less

Note: Metal grains are three dimensional and exist in different sizes. Even if the grains are identical in size and shape their cross sections (seen on the polished surface) will show varying areas depending upon the place that cuts each grain. Therefore no two fields of observation of the structure will be the same. Thus the measurement of grain size cannot be precise but only gives us an idea of the overall size (of the grain distribution) in the metal.

Table 1.1 Data concerning the ASTM austentic grain-size standards*.

ASTM grain-size number	Number of grains per sq in. at 100 X	Calculated diameter of equivalent spherical grain, not magnified			Calculated mean average of cross section of grain, sq in., not magnified
	Mean	Range	Inches	mm	
1	1	0.75 – 1.5	0.01130	0.287	0.0001
2	2	1.5 - 3	0.0080	0.203	0.00005
3	4	3 - 6	0.00567	0.144	0.000025
4	8	6 - 12	0.00400	0.101	0.0000125
5	16	12 - 24	0.00283	0.0718	0.00000625
6	32	24 - 48	0.00200	0.0507	0.00000313
7	64	48 - 96	0.00142	0.0359	0.00000156
8	128	96 - 192	0.00100	0.0254	0.00000078

* From ASTM Standards, Part 1-A, 1946.

CHAPTER 2

Diffusion

Diffusion is a process leading to equalization of chemical composition. In discussing the basic diffusion theories, the concentration rather than chemical potential or activity of the diffusing component is used on the assumption that the solution is ideal.

Fick's first law - steady state of diffusion: The diffusive flux J is defined as the amount of a diffusing species crossing a surface of unit area perpendicular to the direction of flow in unit time, this flux is equal to the product of the diffusivity D and the concentration gradient $\dfrac{\partial c}{\partial x}$ thus,

$$J = -D \frac{\partial c}{\partial x}$$

where

C	=	amount of suvbstance per unit volume
x	=	distance in the direction of diffusion
$\dfrac{\partial c}{\partial x}$	=	concentration gradient
D	=	coefficient of diffusion or diffusivity

The diffusivity has the dimensions of (length)^2/time; using c.g.s units, D is given in cm^2/sec. The coefficient of diffusion is a function of state i.e., varies with temperature, pressure and composition only.

The above equation states Fick's first law and is directly applicable to steady state diffusion with constant diffusivity. When a steady state is reached, flux J at any point along the diffusion path is constant, independent of time or distance. This is illustrated in the Fig. 2.1 showing the diffusion of a gas through for example, a metal diaphragm fixed inside a tube, the wall of which are impervious to the gas.

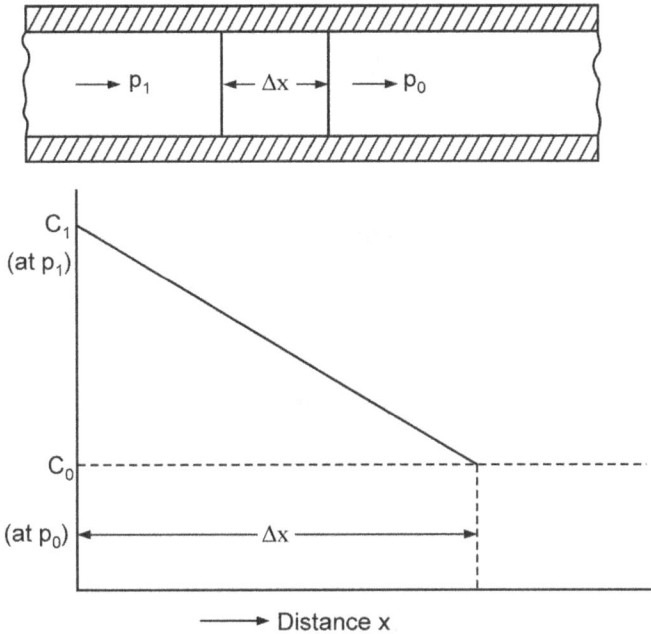

Fig. 2.1

If the partial pressures p_1 and p_0 are kept constant on either side of the membrane and that $p_1 > p_0$ i.e., in terms of gas concentrations $c_1 > c_0$ at steady state, the concentration gradient through the diaphragm is constant

$$\frac{\partial c}{\partial x} = \Delta c / \partial x = (C_0 - C_1) / \Delta x$$

Since the diffusivity is assumed to remain constant for a given temperature, the flux is given by the first law

$$J = -D \frac{(C_0 - C_1)}{\Delta x}$$

Fick's second law: Diffusion in non steady state

In nonsteady state diffusion, the flux changes with distance x. From the first law, this change

$$\partial J / \partial x = -\frac{\partial}{\partial x}\left[D \frac{\partial c}{\partial x}\right]$$

The difference in flux is equal to $-\partial c/\partial T$, negative rate of concentration change. Therefore

$$\frac{\partial c}{\partial t} = \frac{\partial}{\partial x}\left[D\frac{\partial c}{\partial x}\right]$$

If the diffusivity D is independent of concentration (or substantially so under the conditions of experiment), Fick's second law can be written as

$$\frac{\partial c}{\partial t} = D\frac{\partial^2 c}{\partial x^2}$$

The solution to the above equation depends on the geometry and the boundary conditions of the medium into which a substance is diffusing.

(i) Unidirectional diffusion in a semi-infinite medium: Fig. 2.2 shows a cross section of a slab with impermeable side walls. If the length of the slab is large,

Unidirectional diffusion into
a semi infinite medium

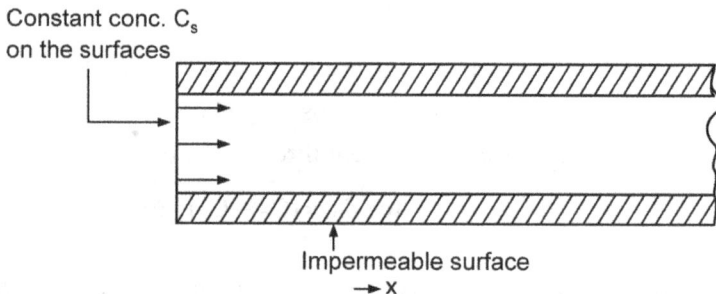

Fig. 2.2 Unidirectional diffusion into a semi infinite medium.

Compared with the distance over which change in composition has occurred due to diffusion, the medium is said to be semi-infinte. At the beginning of the diffusion, the concentration at the surface is instantaneously brought to c_s and maintained constant the throughtout the diffusion time. These boundary conditions are usually abbreviated as

$$C = C_0 \text{ at } t = 0 \text{ and } 0 < x < \alpha$$
$$C = C_s \text{ at } x = 0 \text{ and } 0 < t < \alpha$$

where C_0 = initial uniform concentration

C_s = constant surface concentration

x = distance from the surface S

t = time of diffusion

From the mathematical formula derived for these boundary conditions for constant D, the dimensionless variables have been computed and the values of $\dfrac{(C_x - C_0)}{(C_s - C_0)}$ for any value of $x / \sqrt{Dt}$ can be obtained from the approiate tables. Since these variables are dimensionless, they can be used for any diffusion problem satisfying the boundary conditions given above. For example, if C_0, C_s and D are known, the concentration of the diffusate, C_x at a distance x, can be calculated for any time of diffusion.

(ii) **Diffusion into a slab or cylinder:**   In this case, the diffusion (unidirectional) is considered to take place through the opposite sides of a plate of finite or infinite length or radial diffusion into a cylinder of finite or infinite length. For objects of finite dimensions, comparable with the diffusion distances e.g., thickness of a plate or radius of a cylinder, edges of the plate or the ends of the cylinder must be coated with an impermeable substance to insure unidirectional diffusion. For the boundary conditions, C_0 = initial uniform concentration, C_s = constant surface concentration and constant D, the solutions to the Fick's second law are given in tables, in terms of dimensionless variables. If the diffusivity and the thickness of the plate or the diameter of the cylinder (2*l*) are known, the fractional saturation or desaturation can be evaluated from the dimensioless variables in the tables. The fractional saturation (or desaturation) is given by

$$F = \frac{C_m - C_0}{C_s - C_0}$$

where C_m is the mean concentration of the diffusate.

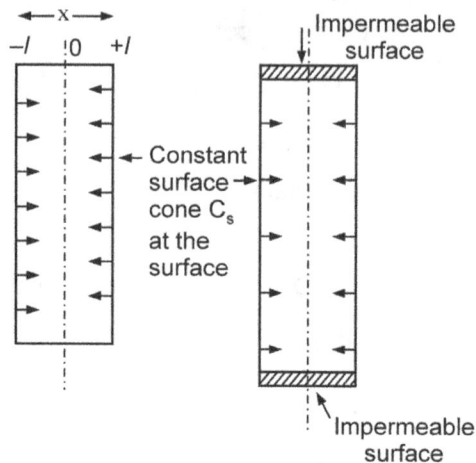

Fig. 2.3 Unidirectional diffusion into plates of thickness $2l$ or into cylinder dia. $2l$ of infinite length or finite length.

INTER DIFFUSIVITY AND INTRINSIC DIFFUSIVITY

If in a binary system, A-B, the diffusing components A and B are identical in mass and size, the rates of transfer of A and B due to random motion across the welded joint ss are equal but of opposite sign. In general, however, the differences in masses and sizes of A and B result in the transfer of A more or less of B. Consequently a hydrostatic pressure tends to build up in the region of solution which contributes least to the volume rate of transfer. This is equalised by the compensating mass flow of A and B together. The resulting concentration profile is as shown in the bottom figure. In order to evaluate the inter diffusion coefficient from this type of concentration profile, the distance x is measured from the reference plane MM. Because of the mass flow during diffusion, the reference plane does not coincide with the original welded junction ss. The amounts of substances diffused on either side of the plane MM are equal to each other but of opposite sign. The choice of this reference plane is first suggested by Matano on an empirical basis and this is referred to as the Matano interface.

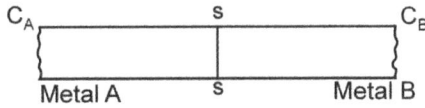

Fig.2.4 Diffusion couple.

If the concentrations C_A and C_B remain unchanged during diffusion experiment, the couple can be assumed to consists of two semi-infinite mediums.

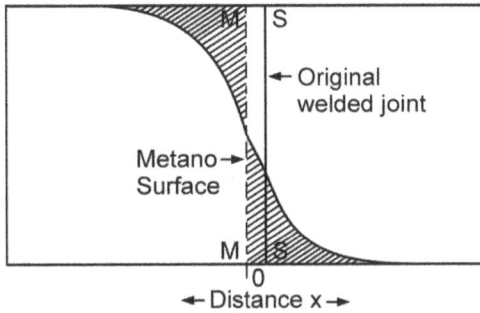

Fig. 2.5 Interdiffusion of A and B through a welded joint.

INTERDIFFUSION OF A AND B THROUGH A WELDED JOINT

If the concentrations C_A and C_B are kept constant on either end of the bar, the interdiffusivity is obtained from the concentration profiles on either side of MM using the dimensionless variables from standard tables.

The mass flow in metallic diffusion couples was first demonstrated by Smigelskas and Kirkendall by showing the movement of inert markers embedded in the metal. This is termed as Kirkendall effect. Therefore in measuring the diffusivities of individual components, the distance must be measured with respect to the position of the inert markers. On the assumption that no volume change occurs during diffusion, the following formula is derived for a binary system A-B.

$$D = N_A\, D_B + N_B\, D_A \qquad\qquad .....(2.1)$$

where   D   =   inter diffusitivity with respect to the Matano interface.

D_A and D_B =   intrinsic diffusivities of components A and B with respect to the inert markers.

N_A and N_B =   atomic or mole fractions $[N_A + N_B = 1]$

The velocity v of the inert marker is related to the intrinsic diffusivities by

$$v = (D_B - D_A) \frac{\delta N_B}{\delta_x} \qquad \qquad .....(2.2)$$

where x is the distance measured, for example from the original welded joint or from the ends of the specimen. Therefore, using inert markers D_A and D_B can be evaluated from the concentration profile and the simultaneous eqns 2.1 and 2.2.

DIFFUSION UNDER A CHEMICAL POTENTIAL GRADIENT

In the absence of an externally applied field, e.g., electrical, gravitational etc., the diffusion occurs when there is a chemical potential gradient in the solution. In solutions obeying Raoult's law, the chemical potential is directly related to the molar concentration. Thus the driving force under isothermal conditions is the concentration gradient. It is for these solutions that the above diffusion eq. no. 2.2 applies.

In nonideal solutions, the chemical potential or activity gradient determines the direction of diffusion and not the concentration gradient. Using Eienstein equation for diffusivity and mobility of the diffusing atoms, the following relation is derived by Darken for the inter diffusivity in terms of self-diffusivities and the activity coefficient for binary systems,

$$D = (N_A D_B^* + N_B D_A^*) \left(1 + \frac{d \log \gamma_B}{d \log \gamma_B} \right)$$

where the asterisk indicates self diffusivity.

The self diffusivity is the intrinsic diffusivity determined from the diffusion of the isotopes of A and B in a diffusion couple of equal chemical potential. In ideal systems obeying Raoult's law or Henry's Law, the ratio d log γ_B / d log N_B is zero. Therefore, intrinsic diffusivity D_i = self diffusivity D_i^* for the i th component.

Since the chemical potential is the driving force, diffusion against the concentration gradient can occur in ternary or multi componrnt systems. This so called up-hill diffusion is demonstrated by the following experiment.

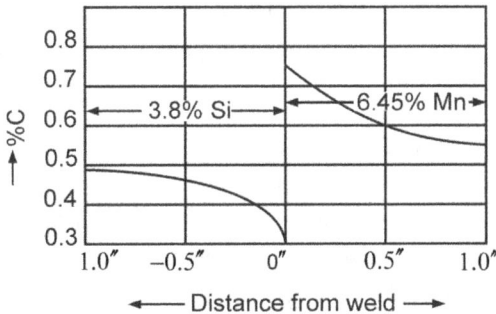

Fig. 2.6 Up hill diffusion of c from low concentration (but high activity in si-iron to high concentration (but low activity) in Mn-iron.

Two steel bars of different compositions are welded and kept at 1050 °C for 10 days. It is seen from the concentration profile that carbon diffused from a low to a high concentration across the welded joint. This is explained simply on the basis of the activity of carbon in the 3.8% si steel being greater than that in the 6.45% Mn steel.

Variation of diffusivity with composition: Although in dilute solutions, the diffusivity can be assumed to be constant, in many alloys diffusivity varies with composition. E.g., C in austenite increases by a factor of about 4 with increasing carbon content within the composition range of 0 – 1.5% C.

Variation of diffusivity with temperature: The logarithem of diffusivity is a linear function of the reciprocal of the absolute temperature, provided the mechanism of diffusion is not attered within the temperature range considered.

CHAPTER 3

Tensile and Torsion Tests

TENSILE TEST

The strength of a material when it is called upon to withstand loads which produce a tensile stress in it, is defined as the tensile strength of the material. A tensile test is the first of its kind of tests. In routine usage, the term "tensile" is omitted and tensile strength is referred to simply as strength.

The conditions of tension are too common to be described in detail. A load which tends to pull apart the two ends of an object is said to be a tensile load. Similarly, readers should be familiar with the terms compression, shear and torsion from earlier education. However, these are illustrated schematically in Figs. 3.1(a) to 3.1(d).

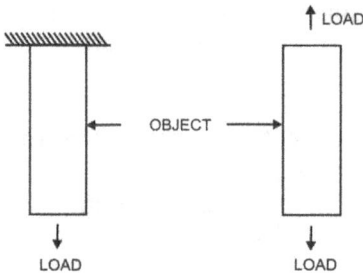

Fig. 3.1(a) Tesion.

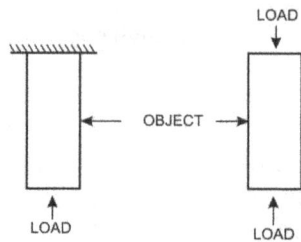

Fig. 3.1(b) Compression.

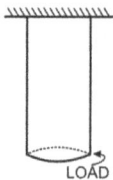

Fig. 3.1(c) Torsion.

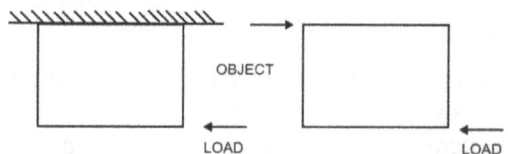

Fig. 3.1(d) Shear.

Before proceeding to the testing it is appropriate to recapitulate some elementary terminology such as load, stress, deformation, strain, etc.

Load

From our standpoint, load is referred to as the weight applied to the body in testing. If a wire is suspended from a nail on a wall and a kilogram-weight is placed in a pan fixed to its bottom end, it is said that the wire is loaded. In this example, the magnitude of the load applied is one kilogram.

In any test, load and its application are important. Load is applied to the extent desired, viz., to the extent the material under test can withstand it before it breaks. The method of application of the load varies depending upon the test performed. In tensile testing the load applied is tensile in nature, i.e., it tends to pull or elongate the specimen. In compression testing, the specimen is subjected to a compressive load, i.e., it is compressed between two opposing loads, and so on. The units for the load are force units and the loads are expressed in weight-Kgf.

Stress

When a body is loaded, forces will be set up inside it. These forces will be in a direction opposite to that of the application of the external load. The magnitude of these internal forces will be such that the effect of the external load is balanced and the body remains in mechanical equilibrium. These internal forces are called stresses. Stress is defined as the internal reaction set up in the solid per unit cross-sectional area. It should be further noted that the internal reaction is proportional to the load applied. At any load, the value of the stress in the solid can be arrived at by dividing the load by the cross-sectional area over which it acts. Stress is express in Kgf/mm^2.

Deformation

When a load is applied on a solid, it tends to change the shape of the solid in its direction. The change affected is called deformation. In the case of tensile loading, the deformation suffered by the object is termed as extension or elongation. Deformation is expressed in the same units as the particular dimension is expressed, i.e., in mm or cm.

Strain

We have seen above that the deformation is the total change suffered by the object in the particular dimension. Strain is the deformation expressed on the basis of the unit dimension. It is customary to express the deformation as strain because the term strain gives a precise measure. In the example illustrated (Fig. 3.2), the strain is l/L. The units for strain are mm per mm. Thus, it is a number.

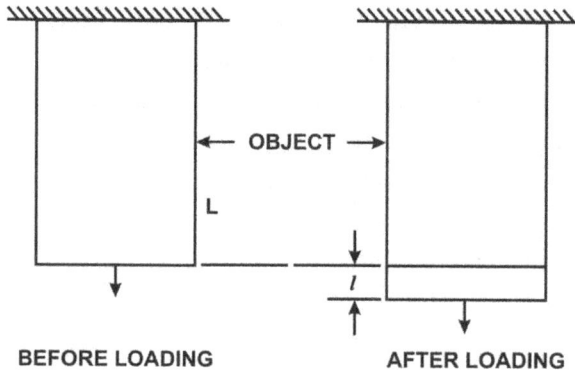

Fig. 3.2(a) Deformation.

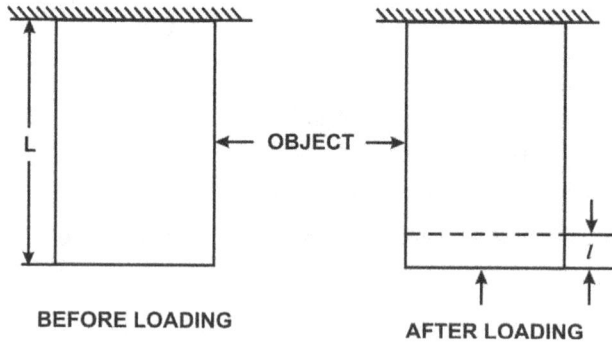

Fig. 3.2(b) Compression.

Stress-Strain Diagram

This is a curve plotted between the stress along the Y-axis (ordinate) and the strain along the X-axis (abscissa) in a tensile test. A material tends to change or changes its dimensions when it is loaded, depending upon the magnitude of the load. When the load is removed it can be seen that the deformation disappears. For many materials this occurs up to a certain value of the stress called the elastic limit. This is depicted by the straight line relationship and a small deviation thereafter, in the stress-strain curve (Fig. 3.3).

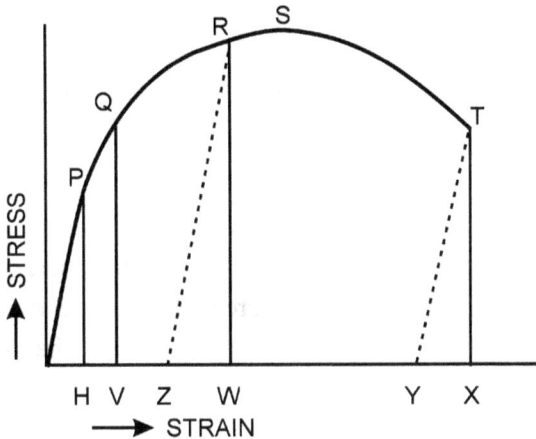

Fig. 3.3 Stress-strain curve.

Within the elastic range, the limiting value of the stress up to which the stress and strain are proportional, is called the limit of proportionality. In this region, the metal obeys Hooke's law, which states that the stress is proportional to strain in the elastic range of loading (the material completely regains its original dimensions after the load is removed). In the actual plotting of the curve, the proportionality limit is obtained at a slightly lower value of the load than the elastic limit. This may be attributed to the timelag in the regaining of the original dimensions of the material. This effect is very frequently noticed in some non-ferrous metals.

While iron and nickel exhibit clear ranges of elasticity, copper, zinc, tin, etc., are found to be imperfectly elastic even at relatively low values of stresses. Actually the elastic limit is distinguishable from the proportionality limit more clearly depending upon the sensitivity of the measuring instrument.

When the load is increased beyond the elastic limit, plastic deformation starts. Simultaneously the specimen gets work-hardened.

A point is reached when the deformation starts to occur more rapidly than the increasing load. This point is called the yield point Q. The metal which was resisting the load till then, starts to deform somewhat rapidly, i.e., yield.

The elongation of the specimen continues from Q to S and then to T. The stress-strain relation in this plastic flow period is indicated by the portion $QRST$ of the curve. At T the specimen breaks, and this load is called the breaking load. The value of the maximum load S divided by the original cross-sectional area of the specimen is referred to as the ultimate tensile strength of the metal or simply the tensile strength.

Logically speaking, once the elastic limit is exceeded, the metal should start to yield, and finally break, without any increase in the value of stress. But the curve records an increased stress even after the elastic limit is exceeded. Two reasons can be given for this behaviour:

(i)   the strain hardening of the material, and

(ii) the diminishing cross-sectional area of the specimen, suffered on account of the plastic deformation.

The more plastic deformation the metal undergoes, the harder it becomes, due to work-hardening. The more the metal gets elongated the more its diameter (and hence, cross-sectional area) is decreased. This continues until the point S is reached.

After S, the rate at which the reduction in area takes place, exceeds the rate at which the stress increases. Strain becomes so high that the reduction in area begins to produce a localized effect at some point. This is called necking.

Reduction in cross-sectional area takes place very rapidly; so rapidly that the load value actually drops. This is indicated by ST. Failure occurs at this point T.

Once the fracture occurs, the load is removed from the specimen. The elastic part of the deformation, i.e., XY is immediately recovered by the specimen. Thus it should be noted that whatever elastic deformation was there in the specimen, it is recovered after the fracture.

We have seen above that once the plastic deformation has started, the cross-sectional area of the specimen starts to decrease. Thus, the actual stress at any moment after the plastic deformation has started is much more. This is called the true stress. Sometimes a curve is plotted between the true stress and true strain. In such a curve the stress value rises and does not show any dipping as in the case of the conventional stress-strain curve shown by the dotted curve in Fig. 3.4a.

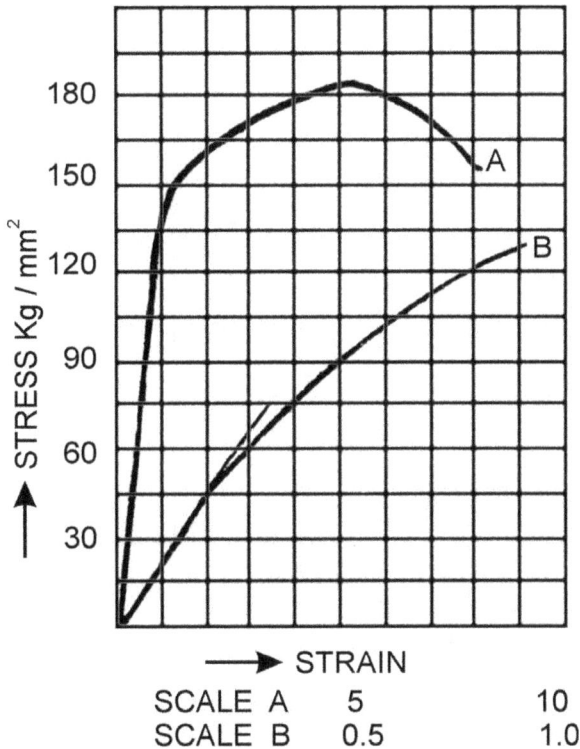

	SCALE A	5	10
	SCALE B	0.5	1.0

A : Engineering stress-strain curve   B : True stress-strain curve

Fig. 3.4(a) Hardened Ni-Cr steel.

Yield Point

Consider the stress-strain diagram of mild steel (Fig. 3.4b). The material has a high proportional limit. When the load is increased beyond the proportional limit, a point is reached when the specimen suddenly starts to deform at a faster rate without any increase in the load. The highest value of stress after which this sudden extension occurs is known as the upper yield point (Yu). The lower yield point (Yl) is the stress which produces a considerable extent of elongation. The upper yield point is depend upon the size and shape of the specimen, its surface finish, and the rate of loading. In routine testing it is the lower yield point which is measured.

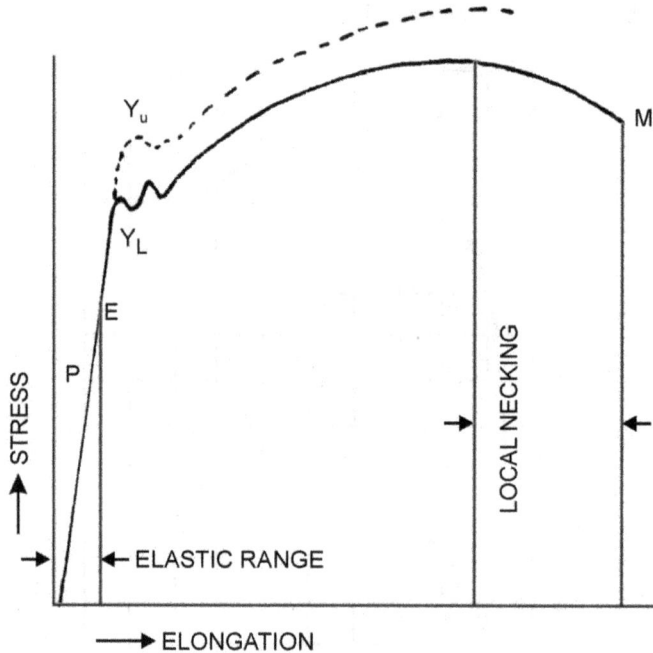

Fig. 3.4(b) Stress-strain curve of mild steel.

The deformation at the yield point is only local in nature. It starts at one point and that region gets work-hardened; so the flow starts again at a region adjacent to the former region. Hardening occurs here too, and the process continues. Thus, the flow is spread throughout the specimen. Each successive work-hardening tends to increase the stress. But the effect is only momentary, and again the stress value falls, due to the flow in the neighbouring region. This is the reason why ups and downs are noticed on the curve in the yield point region (Lüders bands).

As a result, the entire specimen gets work-hardened and the stress begins to rise. Deformation becomes uniform.

Yield Point Phenomenon

It is established that the yield point phenomenon is exhibited by metals and alloys of the body-centred-cubic and hexagonal-close-packed crystal structures which form interstitial solid solutions. Only the four elements carbon, oxygen, nitrogen, and boron form interstitial solid solutions. Thus steel is the most common alloy which exhibits this property. This phenomenon is best explained by the edge dislocation movement. The edge dislocation consists of the compression and tension portions in the slipped and unslipped regions respectively. When it is viewed isometrically, it looks like a pipe pierced through the metal lattice. Calculations show that the diameter of the pipe is much larger than the diameter of the interstitial hole. As a result, the interstitial atoms segregate along the dislocation lines for, in such a configuration, the 'energy' value is the lowest. It is further shown that the screw dislocations also react with the solute atoms. This interaction is very strong when the lattice is non-symmetrically deformed so that a tensile component of the stress is developed. This is called the anchoring of the dislocation. An anchored dislocation will not move and resists the stress. It moves only if a higher stress, sufficient to overcome this anchoring effect of the solute atoms, is applied. When the value of stress reaches this critical value, the anchored dislocations are torn away all of a sudden and rendered free. This is evident by the appearance of the first Lüders band. This triggers the operations in other planes at various other sources. Continued deformation will be evident at a slightly smaller stress, called the lower yield point (Yl). These manifest as the ups and downs in the stress-strain curve and as Lüders bands on the specimen.

It should be noted that the yield point phenomenon is not noticeable above a critical temperature, for the concentration of the solute atoms at the dislocation sites will not be much to make the anchoring effect strong.

Percentage Elongation and Reduction in Area

Before the test is made, the gauge length is marked on the standard specimen (L_o). After the specimen is broken, the two pieces are kept together as if the specimen is not broken at all, with the two fractured surfaces matching each other (Fig. 3.5). The distance between the two

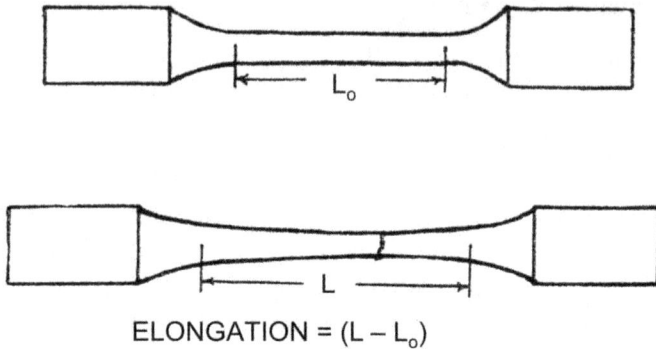

ELONGATION = $(L - L_o)$

Fig. 3.5 Percentage elongation.

gauge length marks is again measured (L). The elongation, and therefrom the percentage elongation, are computed as follows :

Gauge length	$= L_o$
Distance between the gauge length marks after the failure	$= L$
Elongation	$= (L - L_o)$
Percentage elongation	$= \dfrac{\left(L - L_o\right)}{L_o} \times 100.$

Elongation suffered by the specimen owing to the application of the load can be classified into categories, as follows :

(i) The uniform elongation which has occurred until the maximum value of the load S is reached, and

(ii) The local extension produced in the specimen on account of necking. This has occurred after the maximum load S is exceeded.

The uniform elongation is dependent upon the gauge length. As the length of the specimen is increased, the effect of this factor on the total elongation increases. The local extension produced, however, is independent of the gauge length, but it varies with the area of cross-section of the specimen. Thus, it can be said that if the gauge length of the specimen is increased, the effect of necking on the elongation value is decreased*.

---

* It is recommended to use specimens of the same gauge length while studying the ductility behaviours of different metals, from the point of view of the percentage elongation.

This relationship is graphically shown in Fig. 3.6. This is the reason why the gauge length of the specimen is always mentioned while reporting or specifying the percentage elongation. If the failure occurs outside the gauge length marks, the correct value of the elongation cannot be as certained. It is recommended to ignore the test and repeat the test using another specimen.

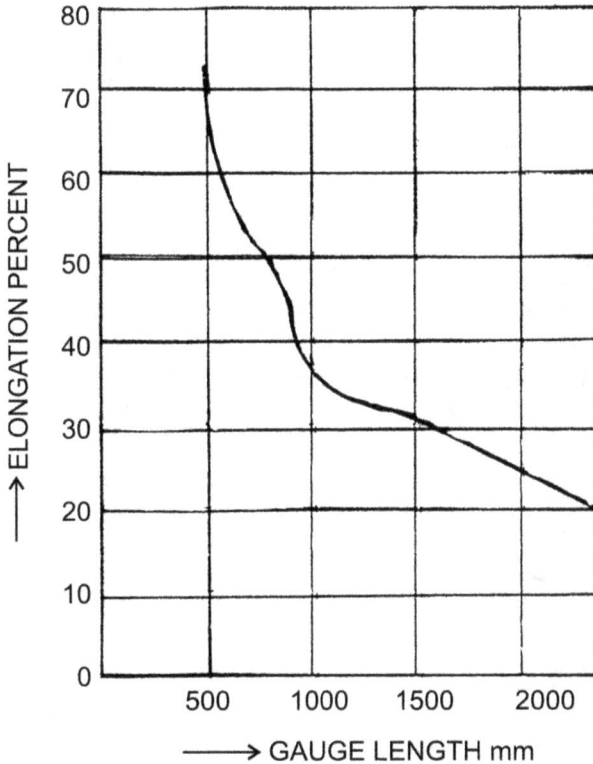

Fig. 3.6 Variation of percentage elongation depending upon gauge length in a low carbon steel.

The test results would be comparable only when the test specimens are geometrically similar, i.e., the ratio of the length (gauge length) to the square root of its diameter is maintained constant (Barba's formula). In India and the United Kingdom the value $L/\sqrt{D}$ is specified as 5.65 as per the recent I.S.O. recommendation.

Percentage reduction in area is the decrease in the cross-sectional area of the specimen up to failure, expressed as a percentage of the original cross-sectional area. Though both percentage reduction in area and elongation indicate the ductility of the material, the former is, in particular, a guide to the formability behaviour of the metal. Though Hadfield manganese steel, stainless steel, copper and aluminium exhibit high values of percentage elongation, the former two possess lower percentage reduction in area. Consequently, they work-harden rapidly and so are not suitable for cold working as aluminium and copper are.

Proof Stress

In many materials including high-carbon and alloy steels and nonferrous alloys, the stress-strain diagram does not indicate a well-defined straight line portion and yield point (Fig. 3.7). Thus, it is not possible to obtain the elastic limit. For engineering design the yielding nature of the material is important. Hence, in the case of such materials a "proof-stress" at a specified strain is calculated. The strain value for the calculation of proof-stress is specified in terms of the gauge length (as a small percentage). Thus, the stress corresponding to a certain allowable amount of plastic deformation is measured and taken in place of the proportionality limit. In checking for acceptance, the material is loaded with the load corresponding to the proof-stress values for 15 secs and the load removed. The material is deemed to have passed the test if it does not show a permanent set, greater than the specified precentage of the gauge length called offset. Usually o.1% permanent set is prescribed for evaluating proof-stress (Fig. 3.8).

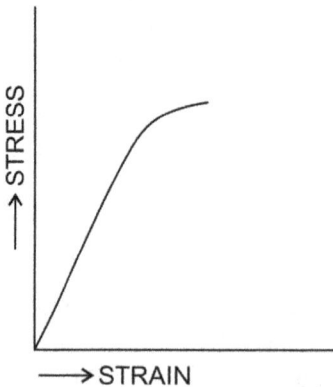

Fig. 3.7 Copper.

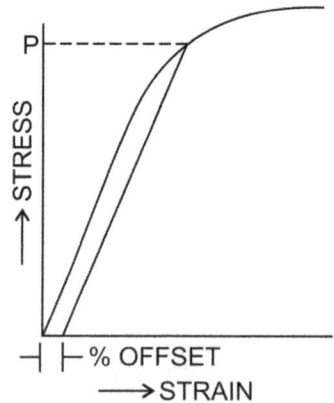

Fig. 3.8 Graphical computation of proof stress.

The most accurate method of determining the proof-stress is only to plot the stress-strain curve very accurately. Usually, a higher load amounting to about 25% of the proof-stress is applied initially. This helps to obtain the elastic portion of the curve at a value higher than zero.

Four-point Method of Calculating the Proof-stress

As it is too laborious and time consuming to plot an elaborate load-extension diagram in routine testing, the four-point method is employed. In this, there is no necessity of plotting any curve. Moreover, it is a fairly accurate method.

This method is aimed at calculating the actual proof-stress, provided the upper and lower proof-stress values are specified. It is the general practice to specify them both. Even if the lower proof-stress is only specified, the upper value may be assumed from our experience with the material. For an easy understanding, the method is explained below with a sketch of a stress-strain diagram (Fig. 3.9).

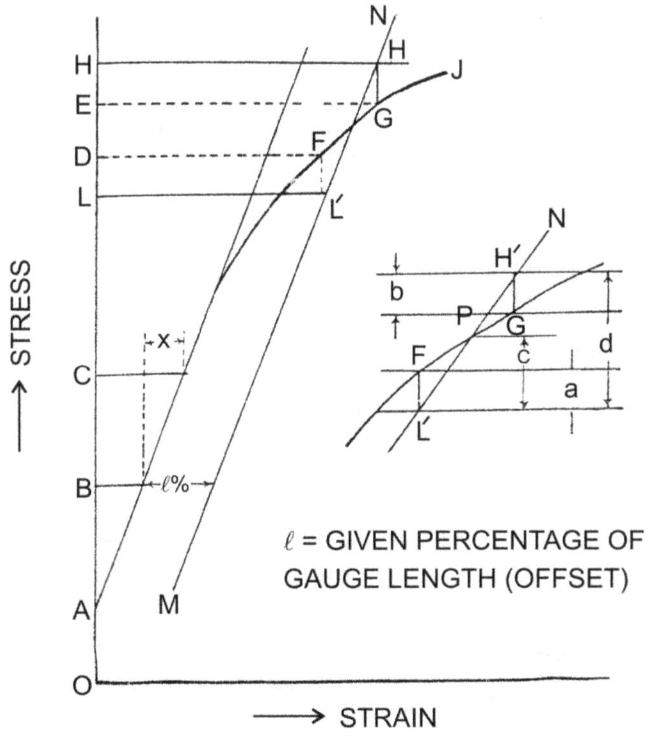

Fig. 3.9 4-point method of determining proof-stress.

The stress-strain curve is shown by *AFGJ*. First, the specimen is subjected to an initial load amounting to *OA* in the diagram. This is about 20% of the* proof-stress of the material. The load is then increased to a higher value, *B*. The extension produced in the specimen is measured. Again, the load is increased to another value, *C*, and the extension measured. In routine testing, this can be directly read by adjusting the extensometer reading to zero at *B*. Let the extension be *x*.

The extension that will be produced at the value of stress amounting to the lower proof stress can be calculated from the relation:

$$\text{Extension (at L.P.S.)} = \frac{AL}{BC} x + l$$

where "*l*" is the specified proof-stress value of the strain expressed as percentage.

The actual stress representing this value of the strain on the diagram *AFGJ* witlh the point *F*.

Similarly, the point *G* can be located by calculating the extension as per the relation $\frac{AH}{BC} x + l$.

Now, if we draw a line MN parallel to the elastic portion of the curve at a distance of *l* from it, we know that the actual proof-stress value can be found out as the point of intersection of the line *MN* with the curve *AFGJ*. Thus, it can be seen that the actual proof-stress of the material lies some where between the two points *F* and *G* on the curve, viz., above *F* and below *G*.

The actual proof stress can be calculated from the relation:

$$c = \frac{ad}{(a+b)}$$

where d = proof-stress range,

a = the amount by which the stress at *F* is above the lower proof-stress value,

b = the amount by which the stress at *G* is below the upper proof-stress value, and

c = the amount by which the stress at the actual proof is above the lower proof-stress value.

---

* From the assumed value.

Thus, the actual proof-stress of the material will be the lower proof-stress value plus c.

For understanding of the above calculation, the proof-stress region of diagram is separately drawn in Fig. 3.9 as an insert.

In actual experiment, the proof-stress can be simply found out as follows :

(i) After properly fixing the specimen in the machine, a load A amounting to 20-25% of the proof-stress of the material is applied.

(ii) The load is increased to about 30-35% of the proof-stress value B. The extensometer is adjusted to zero.

(iii) Again, the load is increased to about 40-45% of the proof-stress value C. The extensometer is read as x.

(iv) The extension at the lower proof-stress value is calculated using the relation $\dfrac{AL}{BC} x + l$ and the load at which that extension is obtained in the specimen is found out. The value of the load is F.

(v) Similarly, the value of the load G which is necessary to give an extension equal to $\dfrac{AH}{BC} x + l$ is found out.

Then, we get

$a = F - $ Lower proof-stress,

$b = $ Upper proof-stress $- G$, and

$d = $ proof-stress range.

(vi) The value of c is calculated from the relation $c = \dfrac{ad}{(a+b)}$ and it is added to the lower proof-stress value. The actual proof-stress is obtained.

THE UNIVERSAL TESTING MACHINE

A photograph of Avery-Denison Universal Testing Machine (model 7113 DCJ) is shown in Fig. 3.10. The top crosshead can be adjusted to three positions for extended tension tests (the left hand side portion of the machine).

The indicator control console contains the pumping unit and the valve gear, the dial indicator and electrical controls. It is connected to the straining unit (right portion of the machine) by flexible pipes and cables. There are two main handwheel controls, one for applying and the other for releasing the load. The loading valve is so designed that at any setting, the rate of oil flow is uniform. When the valve is closed, the load remains stationary long enough for extensometer readings to be taken. The unloading valve is similar and in combination, the two will give all the control needed for applying incremental loads, for applying the loads

Fig. 3.10 Avery-Denison Universal Testing Machine – capacity 40 K.N.
(*Courtesy :* Avery-Denison Ltd., Leeds, U.K.)

quickly, for holding the loads steady and for removing the loads. An autographic recorder (Fig. 3.11) situated at the extreme left of the machine can be used to plot the stress-strain curve during the test itself. A standard tensile specimen to be used on the machine is shown in Fig. 3.12, while three types specimen gripping devices are shown in Figs. 3.13, 3.14 and 3.15.

Fig. 3.11 Autographic Recorder on Model 7108 Universal Testing Machine.
(*Courtesy* : Avery-Denison Lttd., Leeds, U.K.)

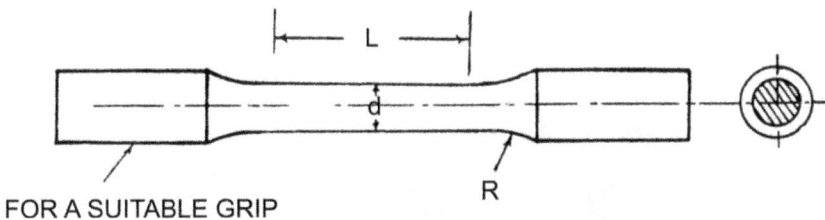

FOR A SUITABLE GRIP

$$L = 5.65 \sqrt{d} \; ; R = d.$$

Fig. 3.12 Tensile specimen. (All dimensions in mm.)

Fig. 3.13 Specimen gripped using threaded grips in the Universal Testing Machine.
(*Courtesy* : Avery-Denison Ltd., Leeds, U.K.)

Fig. 3.14 Photograph showing a plate specimen gripped in the Universal
Testing Machine using pin type of grips.
(*Courtesy* : Avery-Dension Ltd., Leeds, U.K.)

Fig. 3.15 Specimen gripped by wedge grips in a Universal Testing Machine.
(*Courtesy* : Avery-Dension Ltd., Leads, U.K.)

SHEAR AND TORSION TESTS

As briefly explained in the beginning, the shear stress acts parallel to a plane unlike the tensile and compressive stress which acts normal to the plane. The resultants of forces which are parallel, but opposed in direction, act through the sections that are spaced at infinitesimal distances apart. This is an ideal condition of pure shear and the shearing stresses are uniform. A close approximation to this condition is attained in the case of a riveted joint.

If, on the other hand, there is a finite distance separating the two equal and opposite parallel forces, there occurs a tendency to bending in addition to pure shear. A practical example of this condition is that of a beam supported at both its ends (Fig. 1.16). The shearing stress varies from its maximum value at its neutral axis, to zero at the top and bottom layers.

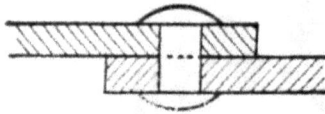

A revised joint illustrating shear.

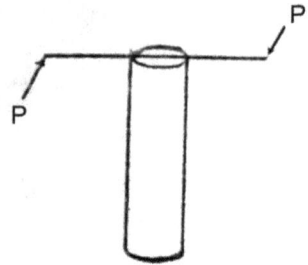

Torsion of a cylinder.

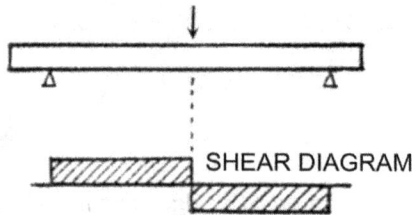

SHEAR DIAGRAM

Fig. 3.16 Beam supported at both ends, showing tendency to bend in addition to pure shear.

When the applied forces are parallel and opposite but do not lie in the same plane as the longitudinal axis of the body, a couple is set up. This couple produces a twisting action about the longitudinal axis of the body and this condition is known as torsion.

CHAPTER 4

Plastic Deformation of Single Crystals

Real crystals do not contain the atoms in perfect symmetry. Some regions will have defects or imperfections. When the deviation from the periodic arrangement of the atoms of the lattice is localized to the vicinity of a few atoms, the defect is called a point defect or point imperfection. If this extends through a microscopic region of the crystal it is called a lattice imperfection. The lattice imperfections can be again divided into line defects and surface or plane defects.

Point defects: There are mainly of three types - the vacancy, the substitutional atom and the interstitial atom. When an atom is found missing from a normal lattice position, a vacancy exists. A number of vacancies can be caused in a pure metal by thermal excitation and these are thermodynamically stable at all temperatures above absolute zero.

$$\frac{W}{N} = \exp\frac{-Es}{KT}$$

where n = No. of vacancies at the absolute temperature T

   N = Total no. of lattice sites or atoms.

   Es = Energy required to move an atom from interior to the surface

and   K = Boltzman's Constant

An atom that is trapped inside a crystal lattice at a point intermediate between normal lattice positions is called an interstitial atom. The presence of an impurity atom at a lattice position is when there occurs a vacancy or the energy requirements are fulfilled. Both interstitial and impure atoms result in disturbances of the periodicity of the lattice.

Line Defects - Dislocations

Dislocation is the most important two dimensional or line defect. This can be imagined to be the region of disturbance.

This can also be said as the boundary between the slipped and unslipped regions of a crystal.

The discontinuation of a row of atoms (in section) is called an edge dislocation.

The shows an edge dislocation with an extra plane of atoms above. This is called a positive edge dislocation. The direction of this extra plane of atoms is the place of the dislocation. Slip occurs at right angles to this place along the plane interacting the dislocation plane at the last extra atom. This is called the slip plane.

The dislocation can also move up or down and this movement in the plane of the dislocation is known as the dislocation climb. This is possible by adding or removing an extra row of atoms to or from the dislocation plane.

Another type of dislocation is the screw dislocation. The upper part of the crystal to the right of AD has moved relative to the lower part in the direction of the slip. No slip has taken place to the left of AD. AD is thus the dislocation line. This is parallel to the slip direction and hence this dislocation is called a screw dislocation. Physical appearance of the disturbed region of the atones also reasonable a screw helix and hence the name.

Slip Deformation

Usually metals undergo plastic deformation by a process called slip. In slip, blocks of the crystals slide over one another along definite crystallographic planes. These planes are known as the slip planes. The existence of slip in metals can be demonstrated by deforming polished specimens and observing their surfaces under the microscope. Steps or equidistant lines can be observed on the polished surface.

Generally the slip plane is the plane of greatest atomic density. The slip direction is the close-packed direction within the slip plane. Since the planes of greatest atomic density are also the most rudely spaced planes, the resistance to slip is generally less for these planes than for others. In the hexagonal close packed metals, the only plane with a high atomic density is the basal plane (0001) and the diagonal axes (1120) are the close packed directions. Similarly in F.C.C. structure (111) planes and (110) directions are closely packed. There are eight (111) planes. But because the planes at the opposite corners of the cube are parallel to each other we can only take them to be four sets. Each plane contains three (110) directions neglecting the opposite directions. In the B.C.C. structure the (110) plane is the highest atomic density plane but it is not superior to several others. Slip in B.C.C. metals is found to occur in (110), (112) and (123) planes while the slip direction is always (111).

Slip in a Perfect Lattice

Slip is supposed to occur by the movement of one plane of atoms over the other. So it is possible to estimate the share stress required for such a movement. Let us consider two planes of atoms subjected to a homogeneous shear stress. The shear stress is assumed to act in the slip plane along the slip direction. The distance between the atoms in the slip direction is to and the spacing between the lattice planes is α. The shear stress causes a displacement in the slip direction between the pair of adjacent lattice planes. The shearing stress is initially zero when the two planes are in coincidence and it is also zero when they have moved one identity distance, so that the atom 1 on the sop plane is over the atom 2 in the bottom plane The shearing stress is again zero when the atoms of the top plane are midway between those of the bottom plane since this is a position of symmetry. Between these positions each atom is attracted towards the nearest atom of the other row so that the shearing stress is a periodic function of the displacement. As a first approximation, the relation between the shearing stress and displacement can be expressed as a sine function.

$$\tau = \tau_m \sin \frac{2\pi x}{b} \qquad \ldots\ldots(4.1)$$

where τ_m is the amplitude of the sine wave and b is the period.

For small values of x, Hooke's law should apply.
So,
$$\pi = Gr = Gx/a \qquad \ldots\ldots(4.2)$$

For small values of b, eq. 4.1 can be written as

$$\tau = \tau_m \frac{2\alpha}{b}$$

$$= \frac{Gx}{a}$$

$$\tau_m = \frac{G}{2\pi} \frac{b}{a}$$

As a rough approximation b can be taken as equal to a. Thus

$$\tau_m = \frac{G}{2\pi}$$

The shear modules for metals is in the range of $10^6 - 10^7$ psi. From the above we in the range of $10^5 = 10^6$ psi. While the actual shear stress required to produce plastic deformation is in the range of 10^2 to 10^4 psi. Even if we give enough allowance for the correction of the sine relation T_m cannot be reduced by more than a factor of 5 from the above prediction. Thus the theoretically calculated value appears to be about 100 times greater than the observed shear strength of metal crystals. It can thus be concluded that a mechanism other than simple body shearing of the crystals is playing a role in slip. That is the presence of dislocations.

Dislocation Movement

In the region of atoms 3, 4, 5 & 6 (shown in the figure), there is an extra plane of atoms. When we consider pairs of atoms say 4 and 5, 3 and 6, they are located symmetrically on the opposite sides of the centre of the dislocation. They encounter forces which are equal distances one half will encounter forces opposing the movement and the other half will encounter forces which assist the motion. Therefore, as a first approximation, the net work required to be done to produce the displacement is zero. The stress required to move the dislocation through one atomic distance is small. (see chapter 1)

The lattice offers no resistance to the motion of the dislocation only when the dislocation lies at a position of symmetry with respect to the atoms in the slip plane.

It can be seen that the movement of a dislocation results in a surface step or slip band when the dislocation reaches the site of the crystal assumed to be a free surface, it produces a shift with respect to the planes on each side of the slip plane. In this case, the shift is equal to one atomic distance because it is a simple cubic lattice. This distance is called Burges vector.

Critical resolved Shear Stress for Slip

The extent of slip in a single crystal depends on the magnitude of the shearing stress produced by the external loads, the geometry of the crystal structure and the orientation of the active slip plane with respect to the shearing stress. Slipping starts to occur when the shearing stress on the slip place in the slip direction reaches an optimum value called the critical resolved shear stress. Consider a cylindrical crystal of cross sectional area A. The angle between the normal to the slip plane and the tensile axis ϕ and the angle which the slip direction makes with the tensile axis is x. The area of the slip plane inclined at ϕ will be A/cos ϕ and the component of the axial load acting in the slip direction is, $\rho \cos \lambda$. Therefore the critical resolved shear stress is given by

$$\tau_R = \frac{\rho \cos \lambda}{A/\cos \phi} = \frac{\rho}{A} \cos \phi \, \cos \lambda$$

The critical resolved shear stress is found to vary with the %age alloy content in an alloy and temperature. If established that the increase in the critical resolved shear stress with alloying will be more, the more is the difference in the sizes of the solid solution forming atoms.

Deformation Mechanism

When the load is applied, the slip planes near the centre of the gauge length rotate so as to align themselves parallel with the tensile axis. The amount of rotation towards the tensile axis increases with the extent of deformation.

As the deformation continues, the rotation of the primary slip system occurs, thus the value of $\cos \phi$, $\cos \lambda$ for this system decreases. Therefore even if the strain hardening is neglected, a greater tensile load should be applied to maintain the value of the resolved shear stress on this slip system. While this is nearing the value of $\cos \phi \, \cos \lambda$ is increasing in another direction (set of planes) which, is being rotated towards 45° to the tensile axis. When the resolved shear stress on this new system is equal to the one over the old system, a new set of slip lines appear on the surface. This process goes on.

Deformation by Twinning

Twinning results when a portion of the crystal takes up orientation that is related to the orientation of the rest of the untwined lattice in a definite symmetrical way. The twinned portion of the crystal is a mirror image of the remainder of the crystal. The plane of symmetry is called twinning plane. In the figure the region to the right is twinned and the one to the left is untwined. The plane of twinning is perpendicular to the plane of the paper.

Twinning differs from slip in several specific respects. In slip, the orientation of the crystal above and below the slip plane is the same often deformation as before. In twinning, there occurs a orientation difference across the twin plane. Slip is usually considered to occur in discrete multiples of the atomic spacing where as in twinning the movements are much less than the atomic spacing. Slip occurs on a particular plane while twinning takes place in a region of the crystal.

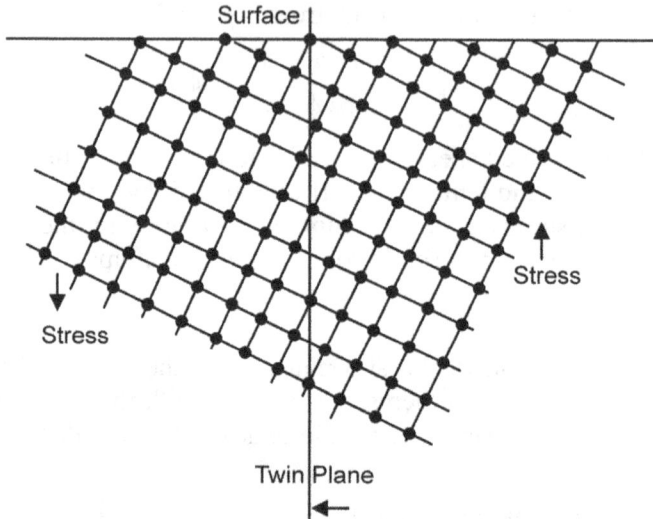

Fig. 4.1 Twinning.

Twins may be produced by mechanical deformation or on annealing a previously deformed crystal. They are known as deformation and annealing twins respectively. Mechanical twins are produced under rapid rate of loading at lower temperatures. Annealing twins result on annealing a crystal deformed by 5-10% for a small time.

Strain Hardening

One of the chief characteristics of plastic deformation of metals is the fact that the shear stress required to produce slip continuously increases with increasing shear strain. The increase in the stress required to cause slip because of previous plastic deformation is known as the strain hardening or work hardening.

Strain hardening is caused by the dislocations interacting with each other and with barriers which impede their motion through the crystal lattice. Hardening due to dislocation interaction is a complicated problem because it involves large groups of dislocations and it is difficult to specify the group behaviour in a simple mathematical way. It is known that the number of dislocations in a crystal increases with strain over the number present in the annealed crystal. Thus the first requirement for an understanding of the strain hardening is the development of a logical mechanism for the generation of dislocations. Frank and Real conceived a mechanism by which a large amount of slip could be produced by one dislocation. Frank Read source provides a method by which the dislocations initially present in the crystal as a result of growth can generate enough dislocations to account for the observed strain hardening.

One of the earliest dislocation concepts to explain strain hardening was the idea that dislocations pile up on slip planes as barriers in the crystal. These pile ups produce a 'back stress' which opposes the applied stress on the slip plane. A zinc crystal is strained to the point 0, unloaded, and then reloaded in the direction apposite to the original slip direction. It can be seen that on reloading, the crystal yields at a lower stress than what it was first loaded. This is because of the back stress developed as a result of dislocations piling up at the barriers during the first loading, aiding the dislocation movement when the slip direction is reversed. Furthermore, when the slip direction is reversed, dislocations of opposite sign could be created at the same source that produced the dislocations responsible for strain in the first slip directions. Since the dislocations of the opposite sign attract and annihilate each other, the net effect would be a further softening of the lattice. This explains the fact that the flow curve in the reverse direction lies below the curve for continued flow in one direction. This lowering of the yield stress when the deformation in one direction is followed by the deformation in the other direction is called the Bashinger effect.

Microscopic precipitate particles and foreign atoms also serve as barriers to dislocation movement. Further, glide dislocations on intersecting slip planes may combine with one another to produce a new dislocation that is not in a slip direction. This dislocation is of a low mobility and can move only by dislocation climb. Such a dislocation is called a sessile dislocation. These act as barriers to the movement of other dislocations.

The back stress is believed to result from dislocations piling up at the barriers. When dislocation intersecting, the active slip plane. The dislocations screwing through the active slip plane is called a dislocation forest. Strain hardening due to the dislocation on cutting arises from short range forces occurring over distances less than 5 to 10 inter atomic distances. This hardening can be overcome by a finite temperature (thermal fluctuations). But strain hardening arising from the dislocation pile ups at the barriers occurs over larger distance and therefore it is largely independent of temperature and strain rate.

The Flow Curve of Metals

When the stress strain curve for a single crystal is plotted with the resolved shear stress vs shear strain, certain generations can be made. The curve can be divided into three stages.

Stage I is the region of easy glide. The crystal undergoes little strain hardening. This implies that most of the dislocations reach the surface and escape. Slip only occurs in one slip system. That is why this stage is also called the laminar flow.

Stage II is nearly straight line, implying that the stress and strain are proportional to each other. This slope of the curve (ratio of the strain hardening modules to the shear modules) is nearly independent of the temperature and the stress.

Stage III is the region of decreasing strain. The process, occurring in this, stage, are called the dislocations which were so far suppressed can now take part. The dislocations piled up during stage II move by cross slip, thereby reducing the internal energy of the crystal.

CHAPTER 5

Fracture Mechanism

When a material is loaded and the stress reaches a value that the material cannot withstand, it breaks. In material testing terminology this is known as fracture. Fracture can be defined as the separation of a material into two or more pieces under stress.

A fundamental study of the process or mechanism of fracture can be made by observing the two fracture surfaces. How the material has ruptured, gives us an idea regarding the behaviour of the material during and prior to the rupture. A brittle material shows no visible plastic deformation at the time of its fracture (Fig. 5.1). The two fractured pieces when joined together will exactly resemble the test specimen before loading. This is not common in metals and alloys but occurs in non-metals and viscous materials like glass.

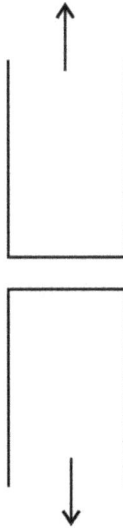

Fig. 5.1 Brittle failure.

53

Metals and alloys show a different type of behavior. A degree of plastic deformation can be found during or prior to the failure. A drastic case is complete necking or drawing-down to too nearly conical surfaces. This is exhibited by very ductile materials like annealed copper. A mixed type of fracture commonly known as the 'cup and cone fracture' is exhibited by a majority of metals and alloys. The cup and cone fracture consists of a core of plastic deformation and an annular ring of brittle failure (Fig. 5.2). Whenever we refer to a ductile failure, it means without saying the cup and cone type of fracture only, as it is the most common in metals.

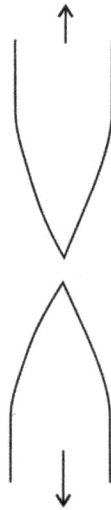

Fig. 5.2 Ductile fracture.

The cup and cone fracture results by a gradual process. The specimen, under the loading conditions prevailing, necks down at the yield stress. When once the necking occurs, the stress distribution which had thereto been uniaxial, changes into a three-dimensional pattern. Small, weaker regions are supposed to be nucleated at this stage and propagate into a minute crack under the influence of the multiple stresses. This crack, generally perpendicular to the direction of the loading, propagates. The crack propagation is governed by the ductility of the material and failure occurs progressively by the plastic deformation reaching its peak in the localized sites. Each region of protrusion (visible in the fracture) can be linked to the complete drawn-down state in a completely ductile fracture

(Fig. 5.3). The crack propagates in this way to the vicinity of the necked surface. Thereafter, the failure will be by cleavage, generally at 45° to the vertical axis. This cleavage region is the brittle fracture region.

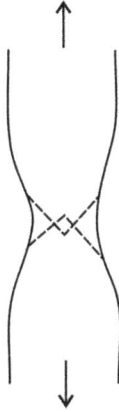

Fig. 5.3 Cup and cone fracture.

COHESIVE STRENGTH

The cohesive strength of metals is very high. A concept of the cohesive strength can be had from the interatomic binding of the solids. Cohesive forces, i.e., the resultant attraction between the atoms, is the difference between the attractive forces between the nuclei of the atoms and the electrons, and the repulsion between the overlapping electron zones. A minima occurs in the energy curve for a particular distance, the interatomic spacing between the atoms. Thus, most of the metals are 'rigid solids'.

If the cohesive strength σ can be plotted against the distance between the atoms, a curve, which can be approximated to a sine curve, is obtained (Fig. 5.4). A tensile load applied tends to increase the separation between the atoms. When plastic deformation starts, the distance between the atoms is increased. Under such conditions, the rate of decrease of the repulsive forces is more than the net forces of balance between the atoms. A condition when the repulsive forces become negligible and the attractive forces also decrease due to the increased separation of the atoms is reached. This corresponds to the peak of the curve, σ max., the cohesive strength of the body.

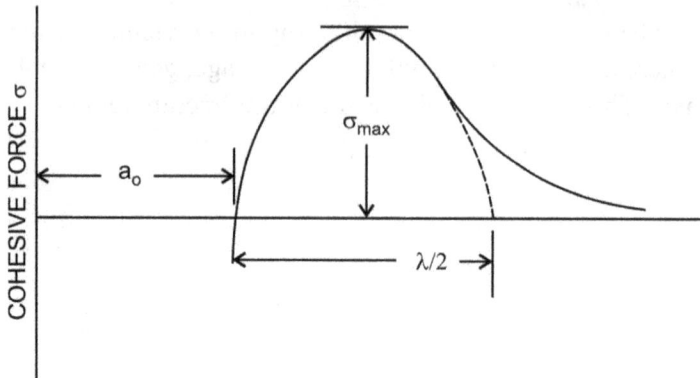

Fig. 5.4 Cohesive force *vs* separation between atoms.

Assuming the curve to be sinusoidal, we get the expression for the curve as,

$$\sigma = \sigma_{max}. \; Sin \; \frac{2\pi x}{\lambda} \qquad\qquad ...(5.1)$$

The work done can be calculated by the area under the curve which can be obtained by integrating the eq. (5.1) between a_o and $(a_o + \lambda/2)$ thus,

$$\text{Work done} = \int_{a_o}^{(\lambda/2 + a_o)} \sigma_{max}. \; Sin \; \frac{2\pi x}{\lambda} \; dx. = \frac{\lambda. \; \sigma_{max}.}{\pi} \; \frac{\cos 2\pi}{\lambda} , a_o \quad .....(5.2)$$

If we imagine that the work done is completely utilized in the formation of a crack and if γ represents the energy required to create a new surface, then we have

$$\frac{\lambda. \; \sigma_{max.}}{\pi} = 2\gamma \qquad \text{or} \qquad \sigma_{max} = \frac{2\pi\gamma}{\lambda} \qquad\qquad ...(5.3)$$

Now we concentrate in disposing of the constants π and λ. For this purpose we approximate the ascending portion of the sine curve (to the tune of $\lambda/4$) to be a straight line. Then, Hooke's law can be applied and can be expressed as,

$$\sigma = \frac{E. \; x}{a_o} \qquad\qquad ...(5.4)$$

where E is the modulus of Elasticity of the material. Differentiating eq. (5.1) with respect to x we get,

$$\frac{d\sigma}{dx} = \sigma_{max}. \frac{2\pi}{\lambda} . \cos\frac{2\pi x}{\lambda}$$

As $\cos 2\pi x/\lambda$ can be taken as unity since the practical deformations are small (x is small), we can omit the term in the above expression, thus,

$$\frac{d\sigma}{dx} = \sigma_{max}. \frac{2\pi}{\lambda} \qquad\qquad .....(5.5)$$

Again, differentiating the eq. (5.4) with respect to x,

$$\frac{d\sigma}{dx} = \frac{E}{a_o} \qquad\qquad .....(5.6)$$

Thus,

$$\frac{d\sigma}{dx} = \sigma_{max}. \frac{2\pi}{\lambda} = \frac{E}{a_o}$$

$$\frac{2\pi}{\lambda} = \frac{E}{a_o . \sigma_{max}.}$$

Substituting for $2\pi/\lambda$ in eq. (5.3) we have

$$\sigma_{max} = \left[\frac{E\gamma}{a_o}\right]^{1/2} \qquad\qquad .....(5.7)$$

This expression is very handy. The values of strength computed from the above equation (theoretical strength) will be many times more than the observed values. However, it is observed that we can go nearer the calculated values by experimenting on single crystal whiskers. This revelation emphasizes the effect of dislocations on the properties of crystal aggregates. Let us examine this in some detail the crystal structure and then crystal behavior under load.

All materials consist of basically atoms. These atoms are arranged in a regular fashion. This arrangement is called a Space Lattice (Fig. 5.5). In this arrangement atoms occupy sites, which are regular and predictable. In other words, the distance between two atoms is equal in all directions. The minimum number of atoms taken as a shape (e.g., a cube), which repeats itself in all the three directions in the lattice is called a 'unit cell'. The shape of the unit cell defines the atomic structure. If the unit cell is a cube, structure of the lattice is defined as cubic and if it is rhombic, the structure is referred to as rhombic and so on.

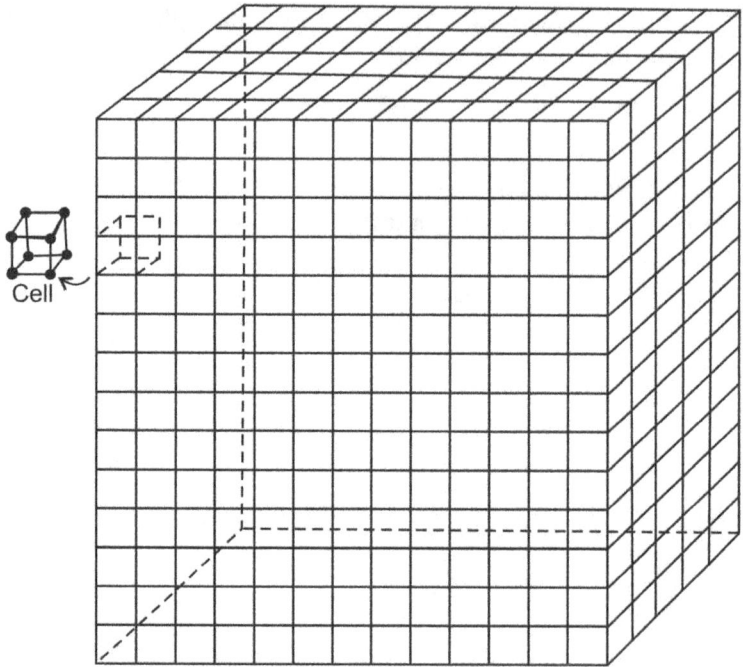

Fig. 5.5 Space Lattice.

The atoms do not occur in a periodic fashion over long distances. Their direction changes after some distance. This changed directions continues for a certain distance and again the direction alters. The existence in a particular direction is called orientation. The region of uniformity is called a crystal and the altered direction in the adjacent region is another crystal. Oftentimes, the changes in the orientations are not abrupt but gradual (over atomic distances). These regions between two crystals are called grain boundaries. This crystalline nature of metals is not visible to naked eye. The metal surface should be polished to a mirror finish and suitably etched with very weak acid solution and observed under a microscope at a magnification of at least 100. The study of metals and alloys under microscopes is known as metallography. Grain (crystal) structures are revealed under a microscope. The regions between two grains, is revealed as lines separating the grains and these are called grain boundaries.

When we observe two specimens under the microscope at the same magnification and when the grains in one appear bigger than the other, we say that the former metal has a coarse grained structure. The structure with more number of grains under view is called *fine grained*. Reporting like this is arbitrary. There are prescribed methods to define the grain size.

As seen above, the lattice orientation in adjoining grains will be different. It is also noted that the change in orientation is gradual. Thus, the grain boundary region will not contain atoms in any stipulated (definable) pattern. These grain boundaries are regions of disturbance. Further, any impurities in the metal will concentrate at the grain boundaries adding to the atomic disturbance.

Grain Structure

So far, we studied about the space lattice taking the example of a cubic cell. The 'unit cell' need not always be cubic. It is defined by its parameters–its edges a, b, and c and angles α (between b and c), β (between a and c) and γ (between a and b). Fig. 5.6 serves to visualize this better. There are fourteen types of space lattices and they fall into seven crystal systems listed in Table. 5.1.

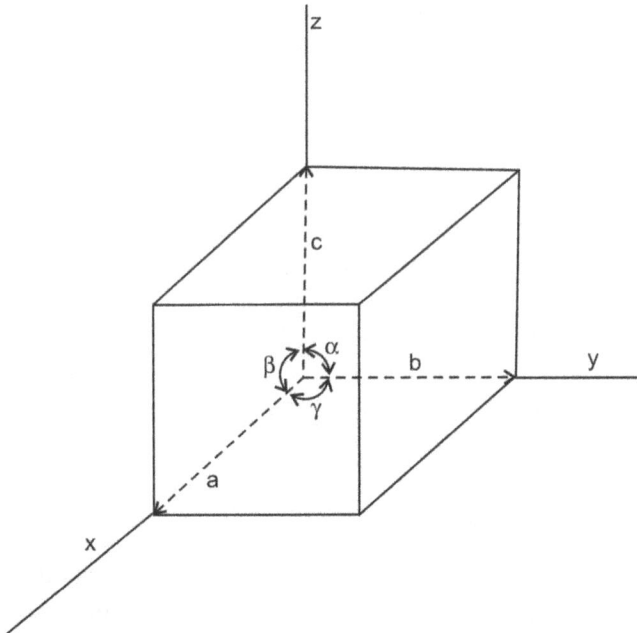

Fig. 5.6 Lattice Parameters.

Table 5.1 Crystal systems*

S.No.	Lattice	Description
1.	Triclinic	Three unequal axes, no two of which are perpendicular $a \neq b \neq c$ $\qquad \alpha \neq \beta \neq \gamma \neq 90°$
2.	Monoclinic	Three unequal axes, one of which is perpendicular to the other two $a \neq b \neq c$ $\qquad \alpha = \gamma = 90° \neq \beta$
3.	Orthorhombic	Three unequal axes, all perpendicular $a \neq b \neq c$ $\qquad \alpha = \beta = \gamma = 90°$
4.	Rhombohedral (trigonal)	Three equal axes, not at right angles $a = b = c$ $\qquad \alpha = \beta = \gamma \neq 90°$
5.	Hexagonal	Three equal coplanar axes at 120° and a fourth unequal axis perpendicular to their plane $a = b \neq c$ $\qquad \alpha = \beta = 90° \qquad \gamma = 120°$
6.	Tetragonal	Three perpendicular axes, only two equal $a = b \neq c$ $\qquad \alpha = \beta = \gamma = 90°$
7.	Cubic	Three equal axes, mutually perpendicular $a = b = c$ $\qquad \alpha = \beta = \gamma = 90°$

*From C.S. Barrett, "structure of Metals", McGraw-Hill Book Company, Inc., New York, 1952.

Again it is our fortune that almost all the metals crystallize into three important crystal structures. These are Face Centered Cubic (FCC), Body Centered Cubic (BCC) and Hexagonal Close Packed (HCP). These are illustrated in Figures 5.7, 5.8 and 5.9.

Face Centered Cubic (FCC)

This structure consists of a cube with eight atoms situated at the eight corners (Fig. 5.7). It also consists of an atom at the centers of each of the six faces of the cube (and hence the name). In this structure, each of the

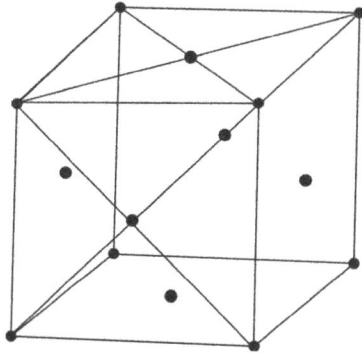

Fig. 5.7 F.C.C Lattice.

corner atoms shared by eight cubes. Each of the face centered atoms shared by two cubes. So we can calculate the atomic density of the structure thus,

In one lattice, 8 corner atoms   ---------- $8 \times \dfrac{1}{8} = 1$

In one lattice, 6 face center atoms   ---------- $6 \times \dfrac{1}{2} = 3$

Total = 4 atoms

The unit cell of FCC structure consists of 4 atoms.

Body Centered Cubic (BCC)

This structure consists of a cube with eight atoms situated at its eight corners (Fig. 5.8). There also exists an atom at the center of the cube. Thus,

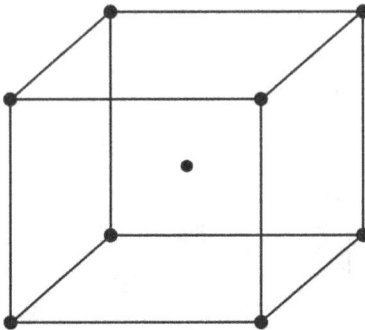

Fig. 5.8 B.C.C. Lattice.

In one lattice, 8 corner atoms   ---------- $8 \times \dfrac{1}{8} = 1$

In one lattice, 1 central atom   ---------- $= 1$

Total = 2 atoms

The unit cell of BCC structure contains 2 atoms.

It can be seen from the above that the FCC cell is more densely packed than a BCC cell. Metals like Chromium, Tungsten, Molybdenum, Vanadium, Sodium and α and δ irons possess this structure. Metals like Aluminium, Copper, Gold, Silver, Lead and γ iron are some examples possessing FCC structure.

Hexagonal Close Packed (HCP)

A hexagonal close packed lattice shows two basal planes which are regular hexagons containing an atom at each of the hexagon corners and one at their geometrical centers (Fig. 5.9). In addition to these atoms, each hexagon contains three atoms at the corners of a triangle in the center of the prism. Their location will be inside the hexagon and inside the alternate rectangular sides.

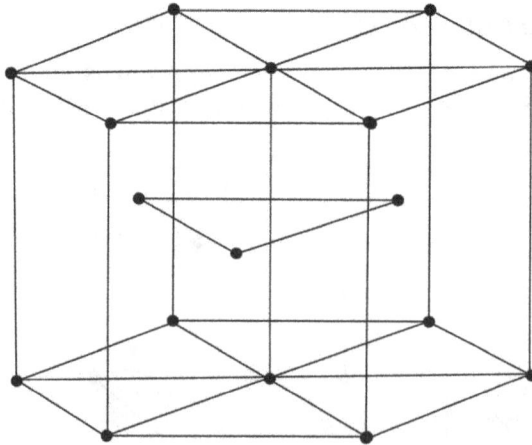

Fig. 5.9 H.C.P. Lattice.

Since each corner atom of the unit cell is shared by eight cells, the basal atoms shared by two and the three atoms inside the hexagon are exclusively of the lattice, the atomic density of the lattice is computed as,

12 corner atoms	---	$12 \times \dfrac{1}{6} = 2$
2 basal atoms	---	$2 \times \dfrac{1}{2} = 1$
3 central atoms	---	$= 3$
Total		$= 6$

Examples of metals that crystallize in this type of structure are magnesium, beryllium, zinc, cadmium etc.

The unit cell of a cubic system can be specified by a single lattice parameter i.e., the side of the cube, a. A hexagonal cell requires the width of the hexagon, a and the height (the distance between the two basal planes), c. The axial ratio $\dfrac{c}{a}$ which is sometimes given. The axial ratio for beryllium is given as 1.58 and for cadmium 1.88.

Polymorphism and allotropy

Polymorphism is the property of the material to exist in more than one type of crystal lattice in solid state. If the change (of structure) is reversible, the polymorphic change is called allotropy. About a dozen metals are known to show this allotropic changes, but the most important and best known is iron. Iron solidifies at $1532°$ C as δ–iron BC C, while cooling further, the δ–iron changes to γ–iron at $1400°$ C which again becomes α-iron at $910°$ C. The γ-iron has an FCC structure while the α-iron has again BCC structure.

Crystal Defects

It is possible to calculate the theoretical strength of a metal by the force required to separate the bond between adjacent atoms. This turns out to be several millions kgf/cm^2. Ordinarily the strength of metals is 100 to 1000 times less. The reason for this lies in the occurrence of defects in the crystal structures of metals.

Real crystals do not contain the atoms in perfect symmetry. Some regions will have defects or imperfections. When the deviation from the periodic arrangement of the atoms of the lattice is localized to the vicinity of a few atoms, the defect is called a point defect or point imperfection. If this extends through a microscopic region of the crystal, it is called Lattice imperfection. The lattice imperfections can again be divided into line defects and surface or plane defects.

Point Defects

These are mainly of three types–the vacancy, the substitutional atoms and the interstitial atom. When an atom is found missing from a normal lattice position, a vacancy exists (Fig. 5.10). A number of vacancies can be

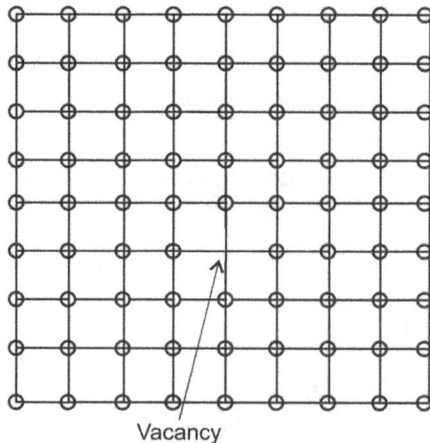

Vacancy

Fig. 5.10 Vacancy.

caused in a pure metal by thermal excitation and these are thermodynamically stable at all temperatures above absolute zero.

$$\frac{n}{N} = e^{\frac{-Es}{KT}}$$

Where,

n = No. of vacancies at the absolute temperature T

N = Total no. of lattice sites of atoms.

Es = Energy required to move an atom from interior to the surface

K = Boltzman's constant

An atom that is trapped inside a crystal lattice at a point intermediate between normal lattice positions is called an Interstitial atom (Fig. 5.11). The presence of an impurity atom at a lattice position is when there occurs a vacancy or the energy requirements are fulfilled. This is called a substitutional atom (Fig. 5.12). Both interstitial and impure atoms result in disturbance of the periodicity of the lattice.

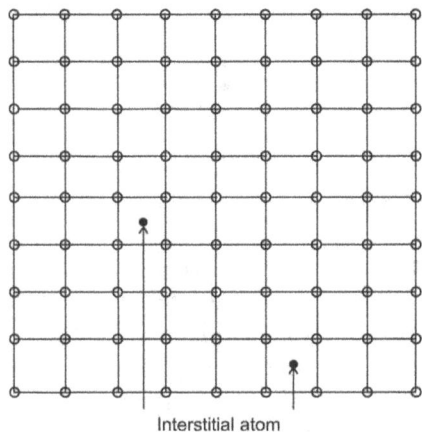

Interstitial atom

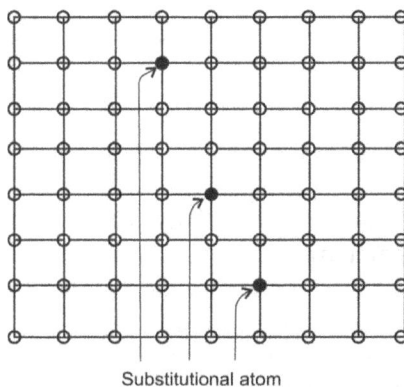

Substitutional atom

Fig. 5.11 Interstitial atom.   **Fig. 5.12** Substitutional Atom

Line Defects/Dislocations

Dislocation is the most important two-dimensional line defect (Fig. 5.13). This can be imagined to be the region of disturbance between two, otherwise perfect lattice structures. This can also be said as the boundary between the slipped and unslipped regions of a crystal. The discontinuation of a row of atoms (in section) is called an edge dislocation. The figure shows an edge dislocation with an extra plane of atoms above. This is called a positive (+ 've') edge dislocation. The direction of this extra plane of atoms is the plane of the dislocation. Slip occurs at right angles to this plane along a plane intersecting the dislocation plane at the last extra atom. This is called the slip plane. The dislocation can also move up or down and this movement in the plane of the dislocation is known as the dislocation climb. This is possible by adding or removing an extra row of atoms to or from the dislocation plane.

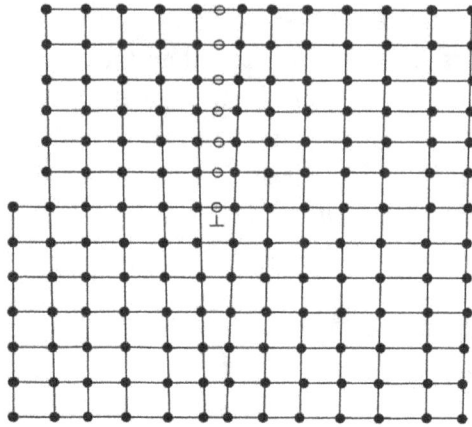

Fig. 5.13 Edge Dislocation.

Another type of dislocation is the screw dislocation (Fig. 5.14). The upper part of the crystal to the right of *AD* has moved relative to the lower part in the direction of the slip. No slip has taken place to the left of *AD*. *AD* is thus the dislocation line. This is parallel to the slip direction and hence this dislocation is called a screw dislocation. Physical appearance of the 'disturbed' region of the atoms also resembles a screw helix and hence the name.

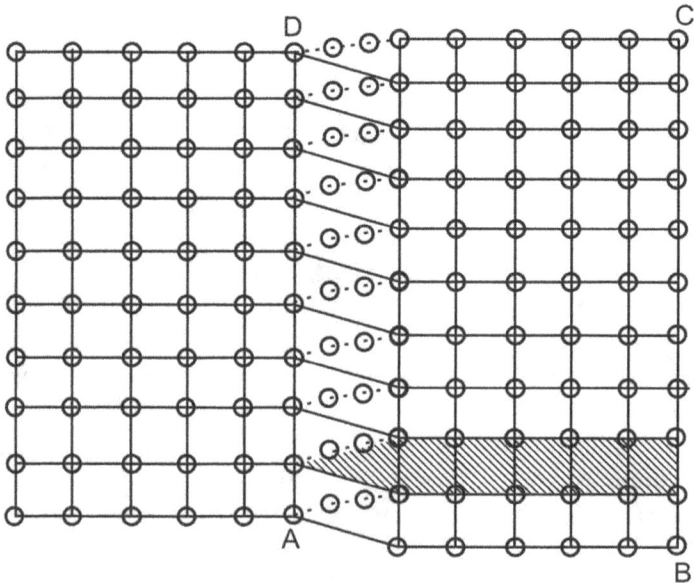

Fig. 5.14 Screw Dislocation.

Metal (or alloy) behavior under stress is dependent upon the effect the stress produces on it. This is the sum of the effects on the individual grains of the metal. So, to have a fundamental concept of the metal deformation, we need to understand the deformation of the single crystal. The slip undergone by the single crystal depends upon the magnitude of the shearing stress produced by the load applied, the geometry of the crystal structure and the orientation of the active slip planes with respect to the shearing stresses. Slip starts when the shearing stress on the slip plane in the slip direction reaches an optimum value called the critical resolved shear stress. This value is really the single crystal equivalent of the yield stress in an ordinary stress-strain curve. The value of the critical resolved shear depends upon the composition and temperature.

Schmid postulated that different tensile loads are necessary to produce slip in single crystals of different orientations can be rationalized by critical resolved shear stress. It is necessary to know from X-ray diffraction, the orientation with respect to the tensile axis of the plane on which slip first occurs and the slip direction of the single crystal tested in tension. Let us consider a cylindrical single crystal of cross sectional area *A* (Fig. 5.15). The angle between the normal to the slip plane and the tensile axis is 'ϕ' and the angle which the slip direction makes with the tensile axis is 'λ'. The area of the slip plane inclined at an angle 'ϕ' will

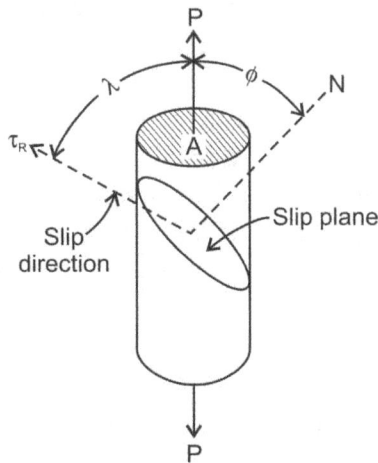

Fig. 5.15 Cylindrical single crystal.

be A/cosφ and the component of the axial load acting on the slip plane in the slip direction is P cosλ. Therefore, the critical resolved shear stress is given by

$$\tau_R = \frac{P \cos \lambda}{\dfrac{A}{\cos \phi}} = \frac{P}{A} \cos \phi \cos \lambda$$

The above equation gives the shear stress resolved on the slip plane in the slip direction. This shear stress is a maximum when $\phi = \lambda = 45°$, so that $\tau_R = \dfrac{1}{2} \left(\dfrac{P}{A} \right)$. If the tension axis is normal to the slip plane $\lambda = 90°$ or if it is parallel to the slip plane ($\phi = 90°$), the resolved shear stress is zero. Slip will not recur for these extreme orientations since there is no shear stress on the slip plane. Crystals close to these orientations tend to fracture rather than slip.

It was experimentally observed that small amounts of impurities in the metal increase the critical resolved shear stress. Addition of alloying elements was found to be of even greater effect. Still greater would be the rise in the value of the critical resolved shear stress in an alloy wherein the solute atoms differ considerably from the solvent atoms in size.

The magnitude of the critical resolved shear stress of a crystal is determined by the interaction of the population of dislocations with each other and with defects such as vacancies, interstitials and impurity atoms. It should be understood that when a positive dislocation moves and encounters a negative dislocation, they cancel out. This stress is of course greater than the stress needed to move a single dislocation yet much lower than the stress required to produce slip in a perfect lattice. On this basis, the critical resolved shear stress should decrease as the density of defects decreases, provided that the total number of dislocations is not zero. When the last dislocation is also eliminated, the critical resolved shear stress should rise abruptly. This rise will be equal to the high value predicted for the shear strength of a perfect crystal.

Most studies of the mechanical properties are made by subjecting the crystal to simple uniaxial tension. In the test, the movement of the cross head of the testing machine constrains the specimen at the grip since the grips must remain in line. As a result, the specimen is not permitted to deform freely by uniform glide on every slip plane along its gauge length (Fig. 5.16). Instead, the slip planes rotate towards the tensile axis since the tensile axis of the specimen remains fixed as in (b). These

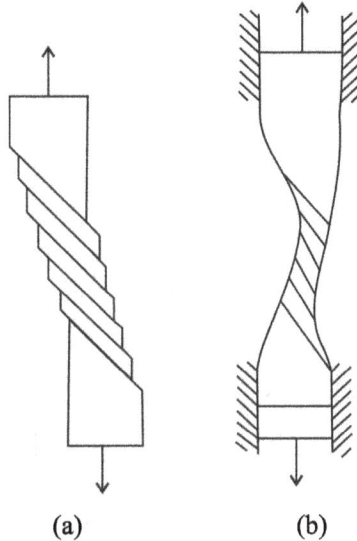

(a) (b)

Fig. 5.16 (a) Tensile deformation of single crystal without constraint
(b) Rotation of slip planes due to constraint.

experimental errors can be compensated by tedious calculations, and it is finally concluded that these can be ignored because of the multitude of grains occurring in the metal specimen under test. This basic concept of deformation at the single crystal level goes a long way in understanding the breakage of the specimen.

FRACTURE OF BRITTLE MATERIALS

Griffith's Theory

Griffith conceived a brittle material to consist of a large number of microcracks. Each of them is capable of producing a stress concentration locally, which equals the σ_{max} of the metal under loads much lower than required ordinarily. If this happens at the region of one microcrack, a crack will form and propagate under favourable conditions. Again, a crack will propagate if the required surface energy for the creation of the two new surfaces (of the crack) 2γ is available. This is obtained by the release of the elastic strain energy hitherto stored in the solid.

Griffith's crack was modelled as an ellipse, the major axis of which is taken as the length of the crack. If the crack is a surface crack, it is taken as half the major axis (Fig. 5.17). The crack propagation stress can be

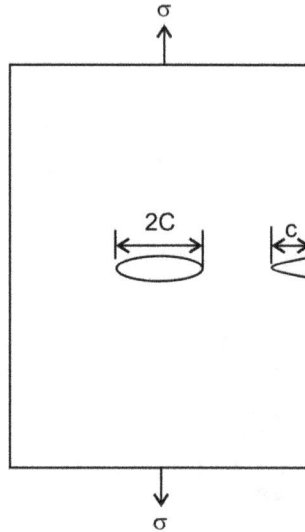

Fig. 5.17 Griffith's crack.

expressed as a function of the size of the microcrack, thus

$$\sigma = \left[\frac{2E\gamma}{\pi c} \right]^{1/2} \qquad \qquad ...(5.8)$$

where c is half the length of the major axis of the ellipse (half the length of the crack).

We can see that if we increase the crack length four times [c is under the square root in the expression (5.8)] the stress value can be halved as they are inversely related.

Orowan's Modification of Griffith's Expression

It sounds somewhat abstract to imagine a metal fracturing without even a trace of plastic deformation. Perhaps this prompted the various investigators to probe the brittle fracture by fractography (examination of the fracture at high magnifications) and X-ray diffraction. It is established that even a brittle fracture has a plastically deformed region, of course, only a few atoms thick. Hence, Orowan introduced a correction factor. This factor takes into account the plastic deformation, in Griffith's expression. The expression is thus modified as,

$$\sigma = \left[\frac{2E(\gamma + p)}{\pi c} \right]^{1/2} \qquad ...(5.9)$$

where p corresponds to the plastic deformation.

CONCEPT OF CRACK EXTENSION FORCE

We have seen while studying Griffith's fracture theory that the crack will form when an energy equal to the value that is required to create two surfaces (2γ) is available. This is given out, as the thereto-stored elastic strain energy, once the fracture starts. Thus, the reversible elastic strain energy is used up in the irreversible process of crack initiation. This is expressed as the 'elastic strain energy release rate' in terms of in lbs per square inch (or mm kgs per mm^2). This is also called the crack extension force. This is a quantity of importance and is claimed to be a material property independent of the size effects but only governed by composition, microstructure, rate of loading and temperature, just as any other material property.

Mathematical expressions show that this term is analogous to the plastic work factor p in Orowan's modification of the Griffith's eq. (5.9).

CONCEPT OF DISLOCATIONS

It is an established fact that dislocations move under conditions of an external load. The dislocation movement is obstructed by :

   (i)   the presence of second phase particles,

   (ii) the grain boundaries, and

   (iii) interlocking by another dislocation.

The obstruction to the movement of dislocations is called a dislocation pile-up. The dislocation pile-up may produce a sessile dislocation, a vacancy, and so on. Detailed studies by eminent workers like Cottrel, Zener and Stroh have established the relationship between grain size, Burger's vector, and the cohesive strength of the metal,

$$\sigma = K.D^{1/2} \qquad \qquad ...(5.10)$$

where σ = fracture stress,

          D = grain size of the metal,

and $\sigma = 2\gamma/n.b \qquad \qquad ...(5.11)$

where σ = crack propagation stress,

          γ = surface energy,

          n = number of dislocations interlocked to give rise to a vacancy, and

          b = Burger's vector.

   With the above study as the background, we can arrive at the following conclusions regarding the fracture behavior of the metals and alloys.

1. The number of dislocations taking part in the dislocation pile-up mechanism leading to the vacancy (crack) formation, controls the metal behavior. If this number (n) is more, the metal behaves as 'brittle'.

2. The grain size of the metal also plays an important role. There appears to be a critical value of the grain size (D) below which the metal fracture will be ductile, and vice versa.

3. There is a temperature range below which the metal fracture will be brittle, and vice versa. This temperature range is called the transition temperature range (correlate to the transition temperature range under impact testing).

4. Higher values of surface energy (γ) tend to promote a ductile fracture. In cases when the surface energy is lowered by the presence of an incongruent precipitate, the metal fracture will be brittle (e.g., hot shortness, H_2-embrittlement).

5. Presence of a notch (c) greatly increases the tendency for a brittle fracture (notch-sensitivity).

References

A.S.T.M., *Special Technical Publication*, 463.

A photographic study of the origin and development of fatigue fractures in Aircraft structures, Ministry of Aviation, U.K., 1961.

Cottrell, A.H., *Trans. Met. Society* AIME, Vol. 212, pp. 192-203, 1958.

Petch, N.J., The fracture of metals, *Progress in Metal Physics*, Vol. 5, Pergamon Press Limited, London, 1954.

Orowan, E., Fatigue and fracture of metals, in the *Symposium at Massachusetts Institute of Technology*, John Wiley and Sons, New York, 1950.

Zenor, C., *Micro-mechanism of fracture in fracturing of metals*, A.S.M., Metals Park, Ohio, 1948.

CHAPTER 6

Creep

Creep is defined as the quasi-viscous flow of the metal. This comparatively occurs at a slow but progressively increasing strain. Creep may be rapid or slow but its rate decreases rapidly as the stress is reduced (this happens at any temperature). But generally the term creep is used to denote the strength of a metal at temperatures much above the room temperature.

High temperature strength of metals is an important requirement. A bearing gets heated up in continuous use, yet it should not soften. The design of moving parts for high speed machinery and many auto and aero engine parts necessitate the employment of materials with good high temperature properties. Rocket technology demands materials with very high and rigid creep requirements. In fact, new alloys have been and are still being developed to meet these challenges of high temperature service requirements.

When a metallic screw cap is jammed we light a match and warm it. The idea behind it is that due to expansion, it gets loosened. Expansion, a property which increases a material's dimensions, will naturally loosen its "joints", i.e., the grain to grain binding strength of the metal is decreased. In other words, the metal becomes plastic or weaker. It is seen in the earlier chapters that the strength of a metal decreases with temperature. This is explained by the greater mobility of the dislocations with the increased temperatures. Further, the atmosphere in which the metal functions, also plays an important part in the creep phenomenon.

Added to the above, the precipitation of hard phases with temperature (ageing) and the grain growth, do exercise their own effects on the metal properties. Thus, the metal properties at the higher temperatures are governed by and depend upon a series of complex variables.

While considering the creep phenomenon from the high temperature serviceability of the metal, strain is as important as the time of exposure. Many metals behave as viscous materials and undergo a time-dependent deformation.

THE CREEP CURVE

A constant stress (load) is applied to a tensile specimen held at a constant temperature. The strain (extension) produced is measured as a function of time and these values are plotted. An ideal creep curve is illustrated in Fig. 6.1.

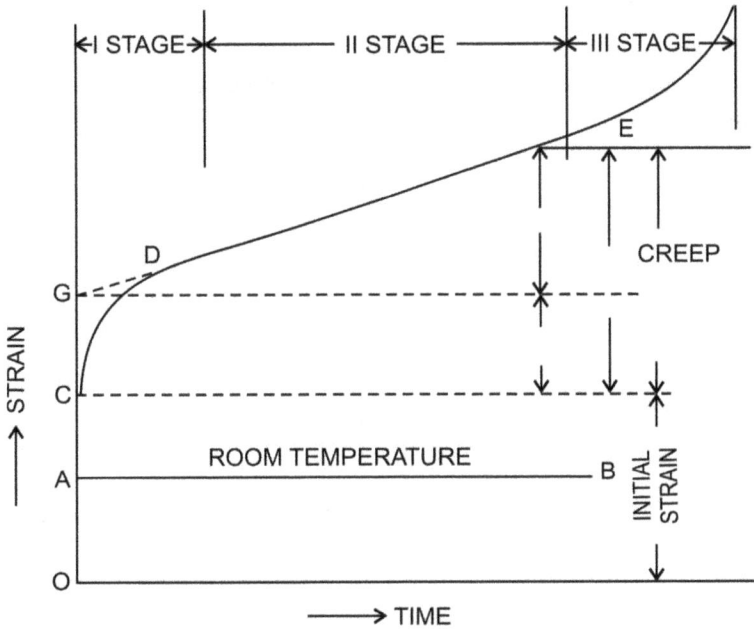

v_0 = Creep rate during II stage
t = Time during II stage
ϵ_0 = Elementary creep
$\epsilon_p = \epsilon_0 + vt$ = Total creep in time t

Fig. 6.1 The creep curve.

The slope of this curve is called the 'creep rate'. When the load is applied to the specimen, there occurs an almost instantaneous elongation. This is denoted by ϵ_0. The creep rate decreases with time during this period and is known as the primary creep. The creep rate later becomes approximately constant in the second stage. The third stage is marked by a rapid increase in the creep rate, which continues until the fracture occurs.

In actual experiment, the third stage is not attained. We have machines with devices which compensate for the necking phenomenon and thereby maintain a constant stress on the specimens throughout the test. Thus a creep curve which appears as an extended second stage curve will be obtained.

Andrade's Concept

Andrade considers that the constant-stress creep curve consists actually of two curves representing two separate creep processes superimposed. The first component is the transient creep which has a decreasing creep rate with time (Fig. 6.2a). The other one is a constant rate creep curve (Fig. 6.2b). When these two are superimposed, the so-called creep curve is obtained (Fig. 6.2c).

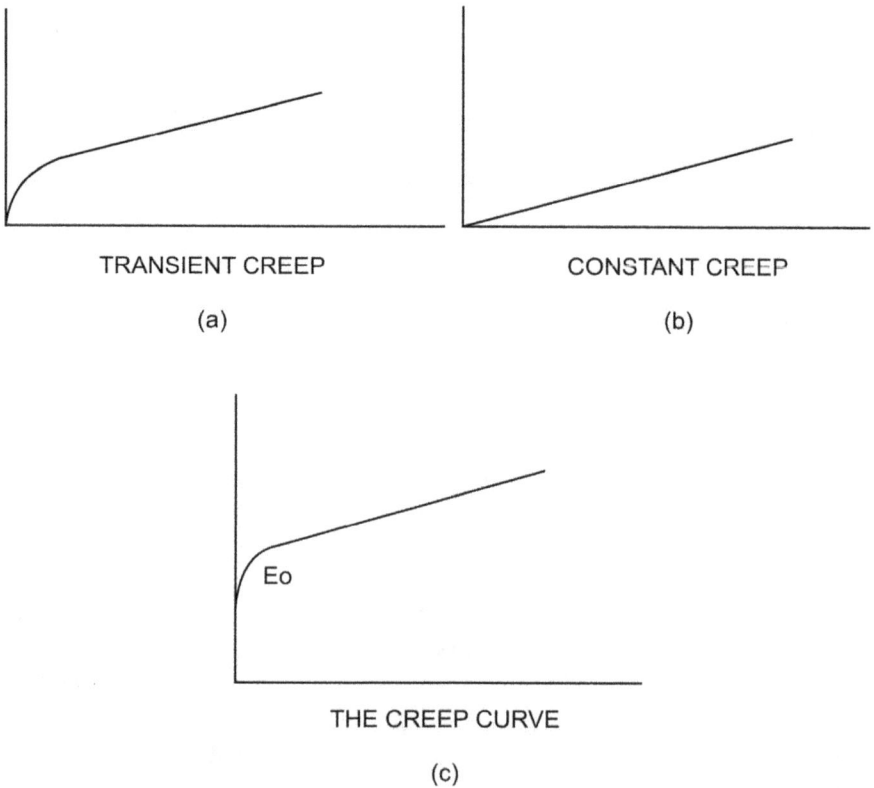

TRANSIENT CREEP

(a)

CONSTANT CREEP

(b)

THE CREEP CURVE

(c)

Fig. 6.2 Andrade's concept.

According to Andrade, the creep curve is represented by the empirical equation –

$$L = L_o (1 + \beta t^{1/3}) \exp Kt$$

where L = length of the specimen at time 't', and

L_o, β and K = constants whose values are empirically determined.

L_o approximates the length of the specimen when the sudden strain produced on the application of the load has ceased. The transient creep is represented by β, a constant. Thus, when $K = 0$ and the above equation becomes

$$L = L_o (1 + \beta t^{1/3})$$

$$\varepsilon = dl / dt = (1/3) L_o \beta t^{-2/3}$$

Again, when $\beta = 0$,

$$L/L_o = \exp Kt$$

$$dL / dt = K L_o \exp Kt = KL$$

or, $K = (I/L) (dL/dt)$

K thus represents an extension per unit length which continues at a constant rate. It is the viscous component of creep.

The transient creep is often referred to as the β-flow and the steady state creep as the K-flow.

In the primary creep stage, mostly the transient creep predominates. A certain deformation of the material occurs and the material gets work-hardened. Thus, its resistance to flow increases. In the second stage creep process, the rate of deformation and work-hardening and the subsequent resistance to flow are nearly equal. A balance will be set up between the opposing processes of strain hardening and recovery. The average creep rate during this stage is termed as the minimum creep rate. The third or tertiary creep state is predominantly the final breaking stage. The reasons for the increase in the creep rate are not yet known but it is supposed that it is due to the necking phenomenon.

EQUI-COHESIVE TEMPERATURE

At room temperature, metals fail by fracture through the grains themselves. We know that the grain boundaries are the regions of high stress concentration and possess no fixed atomic arrangement. Consequently, they are hard and strong. However, this does not hold good in the case of failure at higher temperatures. The grain boundary region undergoes a remarkable degree of 'mobility' leading to a condition of a thin, weak region. The metal in such a state behaves as a material consisting of strong grains embedded in a thin, weak matrix. This is conducive to the grain boundary failure which precisely occurs in creep. It is termed as the intergranular or intercrystalline failure. There exists a temperature for every metal below which the failure will be through the grains, viz., the fracture is transcrystalline, and above which it will be through the grain boundaries, i.e., intercrystalline. This temperature is known as the "equicohesive temperature". At this temperature, the grains and grain boundaries are equally strong.

Equi-cohesive temperature should not be confused with the recrystallization temperature or the transition temperature or even the transformation temperatures of the steels. However, it is nearly equal to the recrystallization temperature for most of the metals. (The equi-cohesive temperature is about 400 °C for plain carbon steels and slightly higher for alloy steels).

GRAIN BOUNDARY FAILURE

Under the conditions of creep, the grain boundary failure may be due to any of two phenomena (Fig. 6.3) :

    (i)   grain boundary sliding, and

    (ii) void formation in the region.

It is seen above that the grain boundary region is weaker than the grain at the higher temperatures. When a load is applied, a sliding action occurs in the region. In such a condition, a crack develops at the regions where three grains meet. This mechanism of crack formation is prevalent when a high temperature and small loading are employed. On the other hand, if the loads are higher and the temperatures lower, voids form at the grain boundary regions which are at right angles to the direction of stress. Gradually these voids grow and coalesce into grain boundary cracks. Many research workers have described the formation of voids phenomenon, the most interesting of which claims that voids form as a result of grain boundary sliding. Investigations into this mechanism are still going on and nothing definite can be put forward, at present.

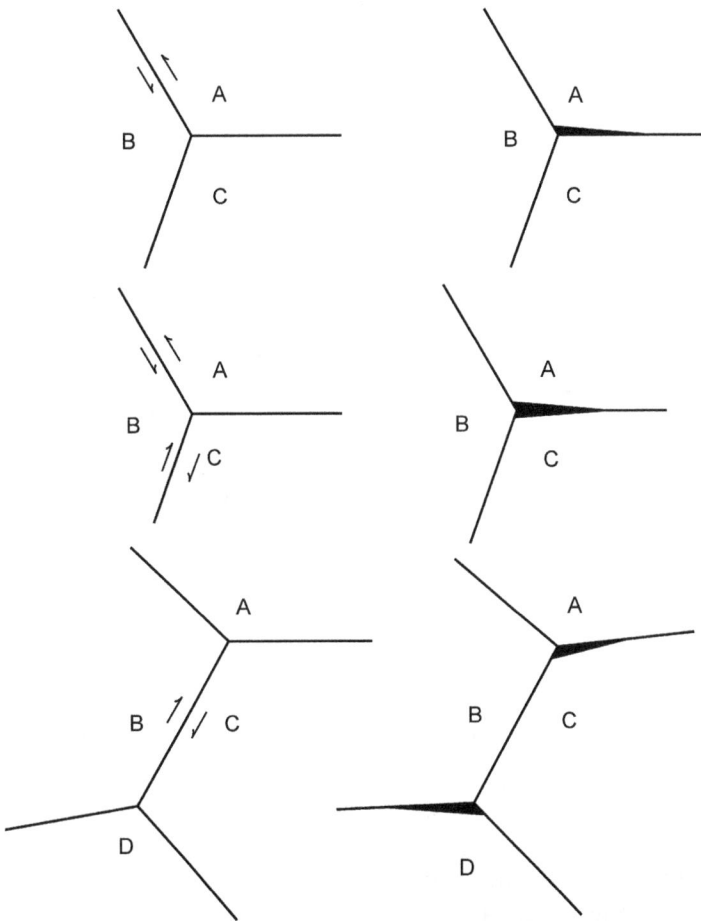

Fig. 6.3 Grain boundary sliding and inter-crystalline crack formation.

(H.C. Chang and N.J. Grant, Trans A.I.M.E., Vol. 206, p. 545, 1956.)

CREEP TESTS

Determination of the creep properties of metals is quite simple. The equipment needed is similar to the one needed for a tensile test. However, the specimen is enclosed in a furnace whose temperature is very accurately controlled and determined. The set-up for creep testing is schematically illustrated in Fig. 6.4.

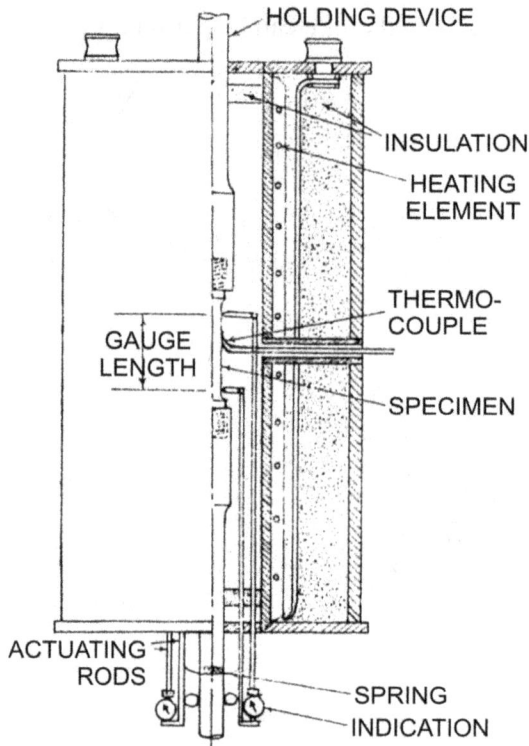

Fig. 6.4 Features of creep testing machine.

Many types of equipment are employed but their basic characteristics are the same. Depending on the design, the furnace accommodates one or more specimens at a time. Whatever may be the unit, the extensometers should be very sensitive and accurate.

While conducting the test, special attention should be paid to the tip of the thermocouple. It should make contact with the specimen under test.

According to the A.S.T.M., the temperature should not vary $\pm 3^\circ$F from the average test temperature. The Load is applied directly and should be measured with an accuracy of 99%.

Standard Specimen

Round specimens of 16 mm, 12 mm and 6 mm (nominal) diameters with gauge lengths equal to four times the diameter are recommended. The larger sizes are preferred because measurement of strain will be easier and accurate.

During the test, the specimen is fixed in the machine and is heated to the temperature of the test in the unloaded condition. After the temperature of testing is attained by the specimen the load is applied; quickly, but without any shock. The initial instantaneous extension is noted, which consists largely of the elastic strain. There after measurements of strain are made at fixed time intervals. At least fifty observations are needed to plot the creep curve. At each observation of the strain the temperature also should be recorded. The average of all these temperature readings should be reported as the actual test temperature.

The Creep Testing Machine

The photograph of an Avery-Denison Creep testing machine (Model Type 48) is shown in Fig. 6.5. The machine has a four-column frame and a loading capacity of 3000 Kgf. The loading is accomplished by a load lever system with a leverage of 40: 1 so that, only a maximum load of 75 Kgf. is used actually. To maintain the perfect axiality of the loading, coplanar knife edged universal joints are provided in the loading system. The standard furnace is wound with Kanthal A in three separate zones and moves vertically on the machine columns to facilitate the specimen replacement. The control cabinet is mounted on the rear main columns and contains the furnace temperature control, variable rheostats for controlling the heat input to the three furnace zones, and an hourmeter to record the time taken by the specimen to rupture.

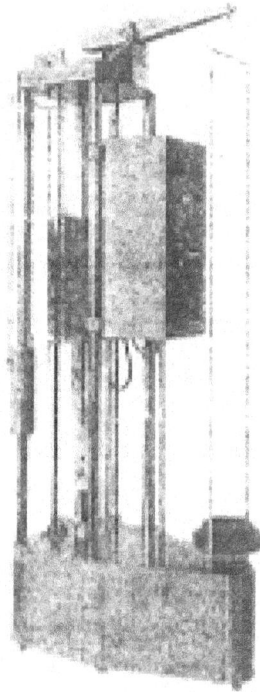

Fig. 6.5 Creep testing machine.
(*Courtesy* : Avery-Denison Ltd., Leeds, U.K.)

Standard specimens that can be used include those up to 16 mm (0.63″ approximately) diam. with gauge lengths of 75,100 or 125 mm, or 6 mm (0.25″ approximately) thick plate specimens. They are fixed with suitable grips made of a high temperature alloy (Nimonic 80A). An optical extensometer or an electronic extensometer (supplied as an accessory) reads the extension. Another important accessory is the controlled atmosphere heating chamber for conducting special tests.

The machine is also supplied as a constant stress rupture machine if required. In this, the standard 40:1 constant load lever system is replaced by a proportional weight lever/cam loading system. The capacity is only of 1000 Kgf., and utilizes a specimen of 25 mm gauge length. The cam is so designed that as the specimen extends, the lever ratio of 10:1 (initial) reduces proportionally with the cross-sectional area to maintain a constant stress on the specimen, over an extension of 12.5 mm (50%).

Interpretation of the Creep Data

Engineering members are designed to last through years and years of service. The number of years of service and the temperature at which they are to operate are the two factors which should be considered in the design. Sometimes the consideration of even these two is not sufficient in design. The permissible plastic strain during the period of service should also be taken into account. The material under consideration is tested, the creep curve is extrapolated to the required number of years on the X-axis and the strain is noted. Once it is less than the maximum permissible strain, the material is deemed to be satisfactory.

Because creep testing is a process which takes a very long time, it is common to test the metal up to the beginning or middle of the secondary creep only. The curve is then extrapolated to the required time period. This process is found to yield good results, for the strains so obtained are always more than the actual strains noted when the test is carried out until such time. The practice gives excess tolerance which is good from the point of view of a sound design.

Another short-cut method to ascertain the high temperature properties is to find out the hot-hardness of the material. It is established that the hot-hardness is linearly related to hot-strength (Fig. 6.6). This method is very useful, especially with brittle materials. However, no idea of the ductility of the material is obtained, which is sometimes absolutely necessary.

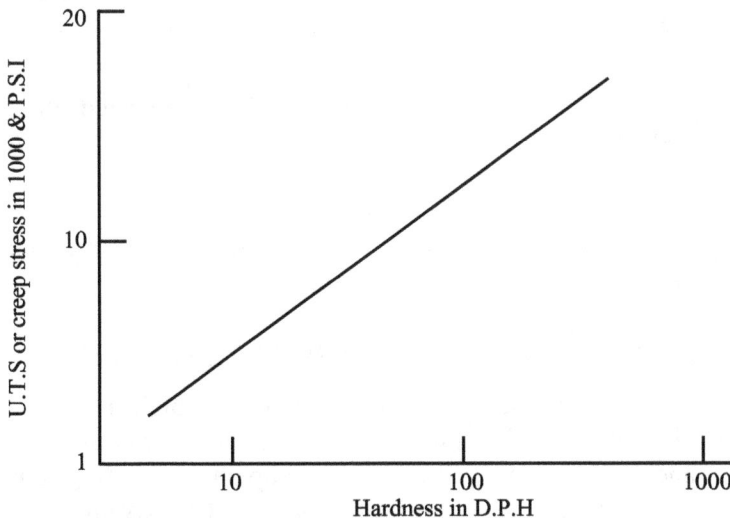

Fig. 6.6

FACTORS WHICH EFFECT CREEP

The following metallurgical factors influence creep properties of steels :

1. Composition
2. Heat treatment
3. Method of steel making
4. Grain size
5. Environment.

Alloying elements which go into solid solution into ferrite Mn, Ni, CO, etc., are found to improve creep properties of steels when the service temperatures do not exceed the recrystallization temperature of the steel. On the other hand carbide forming elements like Mo, W, V, etc., will promote creep properties at temperatures above recrystallization temperature range.

Optimum creep properties are derived when the steel is normalized. Annealed structures are next best. Quenched structures are very inferior.

Electric induction furnace steel is found to possess superior creep properties. Next best is the electric arc furnace steel, whereas, open hearth steel is the worst in this aspect.

A coarse grained steel is found to possess superior creep properties for service above the equi-cohesive temperature whereas a fine grained steel is preferred when operating temperature is below equi-cohesive temperature.

Creep tests on single crystals of zinc have revealed that creep practically stops when they are plated with copper. Again when the copper plating is electrolytically removed, they began to creep at the original rate. Further, the atmosphere in which the metal does service plays a great role. Oxidizing and corrosive atmospheres reduce the creep strength considerably.

SHORT TIME HIGH TEMPERATURE TESTS

Recent advanced technology warrants materials which are loaded and heated up in a matter of seconds. This is especially true in the case of missiles and rockets. Short-time high-temperature testing has thus become a must to ascertain the behaviour of the existing alloys and develop new alloys which withstand these conditions.

Test results are the most dependable when the test is performed under the actual service conditions. The testing consists of heating a special test specimen by its own resistance under a high load. The temperature and elongation are recorded by special automatic equipment. They are plotted to give a curve which resembles an ordinary stress-strain curve, the only difference being that instead of load as in the conventional tensile test, the temperature causes the failure in this case.

Sometimes the temperature required to effect a pre-determined amount of elongation is noted as the yield temperature. The yield temperatures are determined for a number of loads and a curve is plotted between the yield strength and the yield temperature. This curve is a good guide for ascertaining the behaviour of the material at different temperatures or loads (Fig. 6.7). However, it should be borne in mind that short-time high temperature test data is different from the creep data. Both should not be confused.

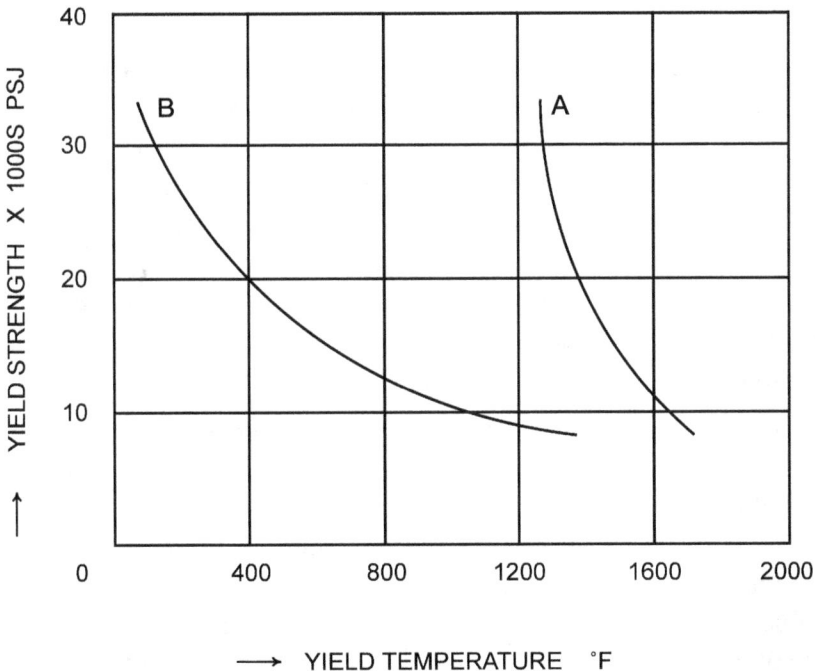

Fig. 6.7 Comparison of short-time high temperature (A) high heating rate (B) yield strength for an annealed 18/8 st. steel (heating rate 800° F/sec).

(Symposium on short-time high temperature testing, A.S.M., 1957)

Relationship between the Creep Rate, Stress and Temperature

It is established that the creep rate is dependent not only on the time of test/service but also on the stress applied/experienced by the metal. An empirical relationship governing the three variables is given by Baunamann as follows :

$$C = f / E. \, T.$$

where C =   creep rate

f =   stress at the particular temperature of test

E =   modulus of elasticity at the particular temperature of test.

and T =   time required to obtain creep equal to the elastic deformation.

HIGH TEMPERATURE MATERIALS

High temperature alloys are developed gradually. Of course, their development was a result of the prevalent needs. In the beginning, alloys suitable for use in steam engines and steam power plants were developed. These were mostly ferritic steels to which chromium and molybdenum were added. These form carbides which disperse throughout the alloy lattices and prevent softening. Typical alloys of this type consist of 0.5 % Mo, 0.1 to 0.3% and 1.25 to 5.0% Cr. These alloys are suitable only up to temperatures of about 600°C. At high temperatures, phase changes occur and the alloys soften.

Later advancements with the advent of jet engines are the austenitic alloys. Nickel based and nickel chromium complex alloys and cobalt based alloys are developed to work up to about 750 °C. In all these alloys the percentage of carbon is very carefully controlled, for if it exceeds an optimum limit, the carbides formed will coalesce instead of being present as finely dispersed particles. The rupture strength is lowered. The austenitic alloys consist of carbon up to 0.1 % (max.), 12 to 25% Ni, 15 to 18% Cr with traces of Cb, Ti, and/or aluminium. These trace additions increase the strength of the alloys by forming compounds with nickel (Ni_3Ti and Ni_3Al) which are stable and hard and get distributed finely throughout the structure. These alloys are harder. The following are some nickel-based high temperature alloys.*

	C	Cr	Ni	Mo	Co	Ti	Al
Inconel	0.04	15.5	76.0	–	–	–	–
Inconel X	0.04	15.0	75.0	–	–	2.5	0.6
Nimonic 90	0.08	20.0	58.0	–	16.0	2.3	1.4
Hastelloy B	0.1	1.0	65.0	28.0	–	–	–
Rene 41	0.1	19.0	53.0	10.0	11.0	3.2	1.6
Udimet 500	0.1	19.4	55.6	4.0	14.0	2.9	2.9

*H.C. Cross and W.F. Simons–Alloys and their properties for elevated Temperature Service. "Utilization of Heat-resisting alloys". A.S.M., 1954.

Some of the cobalt-base alloys are as follows:

	C	Cr	Ni	Mo	Co	W
Vitallium (HS-21)	0.25	27.0	3.0	5.0	62.0	–
X – 40 (HS-31)	0.4	25.0	10.0	–	55.0	8.0

Some complex superalloy compositions are as follows:

	C	Cr	Ni	Mo	Co	W	Cb	Ti	Al	
N–155	0.15	21.0	20.0	3.0	20.0	2.5	1.0	–	–	0.15 N
S–590	0.4	20.0	20.0	4.0	20.0	4.0	4.0	–	–	
S–816	0.4	20.0	20.0	4.0	Balance	4.0	4.0	–	–	3.0 Fe.
K–42B	0.05	18.0	43.0	–	22.0	–	–	2.5	0.2	
Refract aloy 26	0.05	18.0	37.0	3.0	22.0	–	–	2.8	0.2	18.0 Fe.

Note : All the above alloy compositions consist of iron as the balance percentage, except S - 816.

The latest class of the dispersion strengthened high temperature alloys consist of Al_2O_3, ZrO_2 and SiO_2. These oxides are introduced artificially into the alloy matrix to function as the thermally stable phases. Powder metallurgy comes to the rescue of the metallurgist here. After sintering, the alloy is extruded into the required shape.

Based upon the sintering technique, another stride has been made in the development of high temperature alloys. Alloys called cermets are developed by blending ceramic powders such as barites, silicides and carbides with a metal binder. Alloys which stand up to 900 °C have been successfully developed by this technique. Still there is a long way to go in the development of high temperature alloys.

References

Johnson, A.E., Henderson, J. and Khan, B., Complex stress creep relaxation and fracture of Metallic alloys, H.M.S.O., 1962.

Underwood, E.E., J. Inst. of Metals, Vol. 88, pp. 266-271, 1959-60.

Underwood, E.E., Materials and Methods, Vol. 45, pp. 127-129, 1957.

Creep and Recovery, A.S.M., Metals Park, Ohio, 1957.

Garafalo, F., Smith, G.V. and Royle, R., Trans. A.S.M.E., Vol. 78, pp. 1423-1434, 1956.

Sully, A.H., Recent advances in knowledge concerning the process of creep in metals, Progress in Metal Physics, Vol. 6, Pergamon Press, London, 1956.

IS : 3407-1965, Interrupted creep testing of steel at elevated temperatures.

IS : 3408–1965, Non-intercepted creep testing of steel at elevated temperatures.

IS : 3409-1965, Creep stress rupture testing at elevated temperatures.

CHAPTER 7

Fatigue

Like humans, metals too suffer from fatigue. Being humans, we take rest and some food when tired. Metals and alloys do not have that facility and consequently succumb to the fatigue. That is how we can give a concept of the fatigue phenomenon, from a layman's viewpoint.

Fatigue of metals is the property by which they fail at a relatively low value of stress when the stress is repeated. Just as we have seen in the case of impact where the material breaks at a low value of stress (compared to the tensile strength of the material), fatigue failure also occurs at a low stress value. But fatigue is a time-dependent process. An engineering component which is worked continuously for some months or years, whatever the case may be, gives way without any alarm and all of a sudden.

Though the phenomenon is known for over a century, it attained due importance only after the World War II. The value of a fatigue failure can best be appreciated if the reader imagines himself in the position of a lone driver on a highway and the mainshaft of his car is fractured. In the case of an automobile it is only a matter of inconvenience whereas in air and sea travel it is a question of hundreds of lives.

Repeated Loadings

All the engineering components are subjected to loads during service. These loads need not necessarily be steady. In a majority of cases, their magnitudes vary. The material fails, once these fluctuating stresses reach sufficiently high values. The value of this stress may not be anywhere near the static strength of the material. This limiting stress which causes fatigue failure is called endurance limit.

The variation in the stress is generally expressed in two ways :

1. By stating the range of stress that is to be experienced by the metal. This indicates the maximum and minimum values of the stress that the metal will experience. The ratio between the two is known as the 'range ratio'.

2. When the stress is fluctuating, a mean value of the stress should be specified along with the alternating stress to be superimposed to produce the given variation in the stress condition. Thus, the maximum values of the stress can be calculated by superimposing the stress range over the mean value.

A classification of the types of the repeated stresses is given in Fig. 7.1. This classification is based on the report of the ASTM research committee on fatigue of metals. In all these classifications, however, the nature of the stress, viz., tension, compression, torsion, etc., should be specified.

A common type of engineering design, in which each of its components should be designed taking into account its fatigue strength, is a road or rail bridge. When a lorry or the rail is moving on the bridge, the structural members are subjected to repetition of the loads. Further, at any single point on a railway bridge, the loads become cyclic between the passage of the wheels across the point. In many other cases, stresses are quite severe and more frequently repeated. The main shaft and other part of a steam or oil engine, springs, wire ropes, crane hooks, car-axles, rails, railway axles, etc., come under this category. Moreover, the loading experienced by them is mostly cyclic.

Fatigue Strength and Endurance Limit

In a fatigue test, the value of stress at which the metal fails is called the fatigue strength. But in practice, the components are never designed to develop that value of stress which is equal to the fatigue strength by employing factors of safety in design. A limiting stress calculated from the endurance limit is always taken into account. The endurance limit is determined for a particular number of cycles of stress.

Fatigue Test Loading Conditions

Of the types of repeated stresses described in Fig. 7.1, the totally reversed stress is the most severe one. That is why fatigue strength of metals is generally determined under completely reversed bending conditions. Thus, the values or data are always taken to be that of completely reversed bending test only, unless otherwise specified.

TYPES OF REPEATED STRESSES*

TYPES OF STRESS VARIATION		RANGE RATIO NOMENCLATURE		MEAN STRESS NOMENCLATURE	
DESCRIPTION	DIAGRAM	MAX. STRESS	RANGE RATIO	MAX. STRESS	ALTERNATING STRESS
STEADY STRESS σ_1		σ_1	$\frac{\sigma_1}{\sigma_1} = 1.0$	σ_1	0
PULSATING STRESS BETWEEN σ_1 AND σ_2	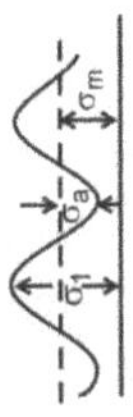	σ_1	$0 < \frac{\sigma_1}{\sigma_2} < 1$	σ_m	$\pm \sigma_a$
PULSATING STRESS BETWEEN σ_1 AND 0		σ_1	$\frac{0}{\sigma_1} = 0$	σ_m	$\pm \sigma_a$
PARTLY REVERSED BETWEEN σ_1 AND (−) σ_2 WHERE $\sigma_2 < \sigma_1$ AND OF OPPOSITE SIGN	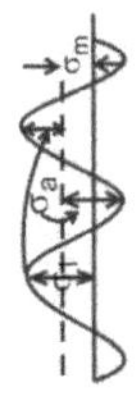	σ_1	$-1 < \frac{-\sigma_2}{\sigma_1} < 0$	σ_m	$\pm \sigma_a$
COMPLETELY REVERSED STRESS BETWEEN σ_1 & σ_2 WHERE $\sigma_1 = \sigma_2$	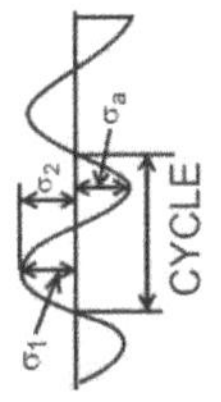	σ_1	$\frac{-\sigma_2}{\sigma_1} = -1.0$	0	$\pm \sigma_a = \sigma_1$

NOTE : $\sigma_m = (\sigma_1 + \sigma_2)/2$ & $\sigma_a = (\sigma_1 - \sigma_2)/2$
*From Proceedings A.S.T.M., Vol. 37, Pt. I, 1937.

Fig. 7.1 Types of repeated stresses.

In general design, the endurance limit is taken as a fraction of the tensile strength. For steels this value lies somewhere between 0.4 to 0.6. It is lower than 0.4 in the case of non-ferrous metals and alloys. The supported beam. In the cantilever type again, the load may be applied at a single point or halved and applied at two places, constituting a couple. When the load is applied at a single point, there is a tendency for the bending moment to concentrate near the fixed end or chuck. This is minimized by giving a suitable taper to the specimen. On the other hand, when the loading is done at two places, a larger volume of the test specimen is brought under the influence of the load (see Figs. 7.2 and 7.3).

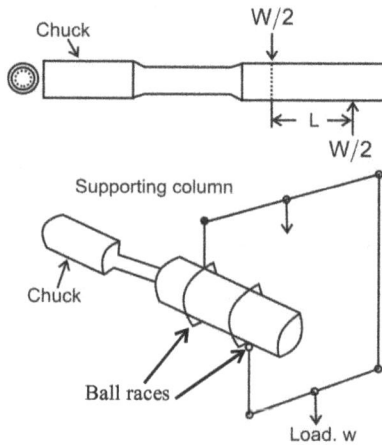

Fig. 7.2 Two-point loading rotating beam fatigue testing.

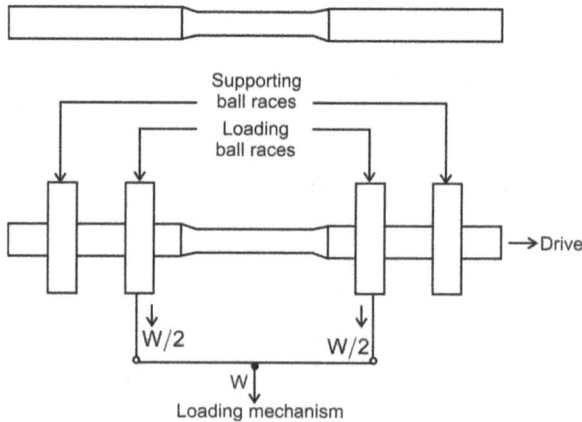

Fig. 7.3 Four-point loading rotating beam fatigue testing.

The simply supported beam type loading is made use of in the Simplex rotating beam machine. The test specimen has an effective diameter of 7.62 mm (0.3"). The long shoulders of the specimen are for mounting it on two sets of ball races, one on each side. The outer races serve as the supports and the inner ones load the specimen [see the line diagram (Fig. 7.3)]. The specimen is rotated by an electric motor. The Baldwin Sonntag rotating bend fatigue testing machine also operates on the same principle but it has a higher capacity. It can test specimens up to 1" (25 mm) effective diameter.

A large variety of fatigue testing machines are in the market. A thorough review of the various machines is given by Forrest.[2] Special machines are developed by national testing laboratories or such bodies or large industries in many countries for special and specific fatigue testing purposes of paramount importance. The National Engineering Laboratory and the London Transport Board (both U.K.) have such huge machines for bend fatigue testing of specimens as big as railway axles (Fig. 7.4).

Fig. 7.4 Heavy duty fatigue equipment (specially designed for a research project).

(*Courtesy* : London Transport Board, Chiswick, U.K.)

Similarly the Moscow Central Scientific Research Institute of the U.S.S.R., Ministry of Transport, have machines capable of testing axles up to 12″ (300 mm approx.) diameter.

Summarizing, the various types of fatigue testing machines incorporate the following:

1. A means of holding the specimen suitably.

2. A means of loading and measuring the load.

3. Provision to rotate the specimen and measure the number of revolutions.*

4. A means of disconnecting the revolution measuring device or putting off the machine itself, once the specimen is broken.

Machine designs differ in respect of the first two features. The specimen may either be loaded as a beam or cantilever. Again, the loading may be simply by mounting weights onto a weight holder or by hydraulic means or attracting and leaving a temporary magnetic head by means of a strong and soft electromagnet (The Amsler's Vibrophore) or by means of an eccentric or crank.

A fatigue testing machine manufactured by M/s. Avery Denison Limited, Leeds, U.K., is shown in Fig. 7.5. The grips for use on combined bending and torsion actions are separately shown in Fig. 7.6. The standard test specimen for use on the machine is shown in Fig. 7.7.

---

*Note : A recent feather in the cap of fatigue researchers is the work of the British Non-ferrous Metals Research Association. They developed a new design for bend fatigue testing of metals at elevated temperatures. In this, the specimen remains stationary and the load is rotated. The stationary specimen facilitates an easy access for temperature measurement.

Fig. 7.5 Laboratory type fatigue testing machine.
(*Courtesy* : Avery-Denison Ltd., Leeds, U.K.)

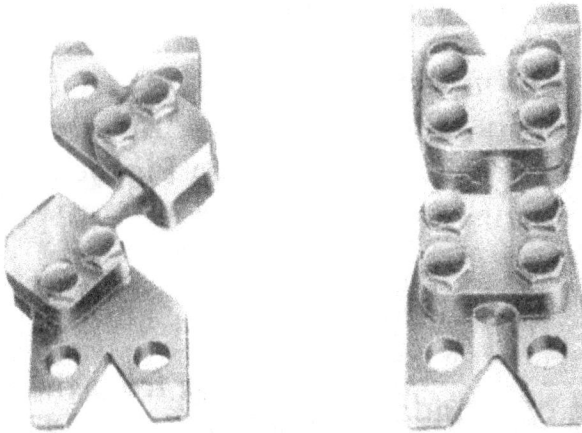

Combined bending and torsion   Bending

Fig. 7.6 Grips for bending of round specimens and grips for combined
bending and torsion of round specimens.
(*Courtesy* : Avery-Denison Ltd., Leeds, U.K.)

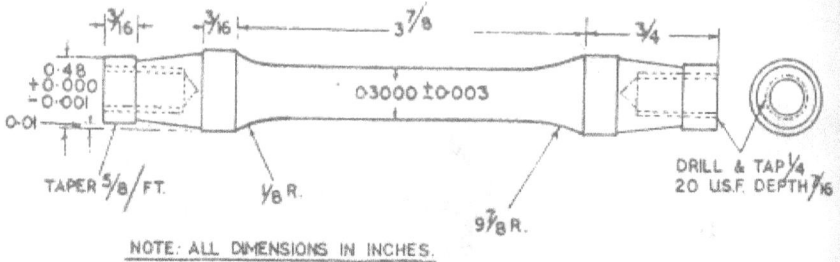

Fig. 7.7 Standard specimen for rotating beam test.
(From *Metals Handbook*, A.S.M., 1948).

TEST PROCEDURE

A number of standard test specimens are made from the metal under test. The first specimen is tested at a high value of load. The number of revolutions the specimen experiences before fracturing is noted on the counter. Another specimen is fixed in the machine and this time, the load is slightly decreased. The number of revolutions that are indicated in the counter this time, will be more than that in the previous case. The other specimens are also broken one after the other, each time reducing the load, and noting the number of revolutions. The procedure is continued until a value of stress is reached when the specimen does not fail, say even after 10 million revolutions.

When the specimen is rigidly fixed at one end and loaded at the other, it bends. In such conditions, the top half of the layers in the specimen experience a sort of a stretching action while those in the bottom half experience compression. In Fig. 7.8 (b), the layers experiencing stretching or tension stress are white while those experiencing compression are shaded. Once the specimen turns half a revolution, it literally comes into an upside down position with respect to its initial condition. In such a condition the stress distribution in the specimen also changes exactly in an opposite fashion. That is, those regions which were under tension become compressed and vice versa [Fig. 7.8 (c)]. Further rotation through 360° brings the specimen back to its initial position. Thus, during one complete revolution, the stresses in the specimen are completely reversed once. It means that, by measuring the number of revolutions the specimen is subjected to before failure, we actually measure the number of reversals of the load.

(a)

LOAD NOT APPLIED

(b)

LOAD APPLIED

(c)

LOAD REVERSED

S

N

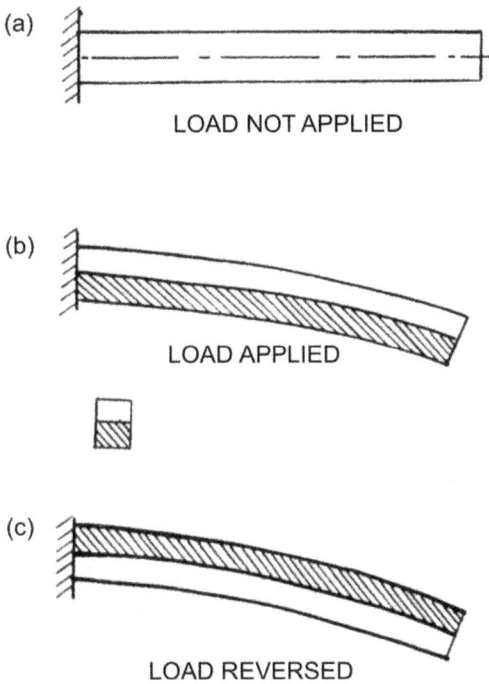

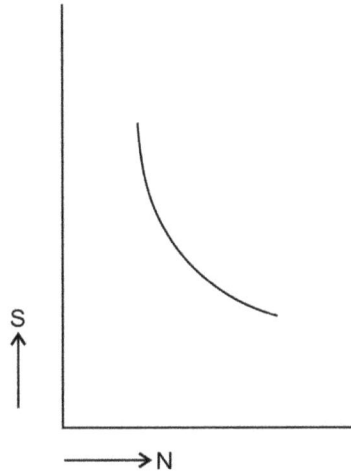

Fig. 7.8 Stress distribution in a cantilever
loaded and rotated.

Fig. 7.9. S-N diagram.

From the data thus obtained, the stress, number of reversals of the
stress diagram, commonly referred to as the S-N diagram, is plotted
(Fig. 7.9). The stress value is plotted on the Y-axis and the number of
reversals of the load, taken on a logarithmic scale, on the X-axis. The
stress value at which the curve becomes nearly horizontal, is taken to be
the endurance limit of the metal.

THE FATIGUE FAILURE FRACTURE

A typical fatigue fracture surface is shown in Fig. 7.10. The fracture is characteristic and easily recognizable because of the following:

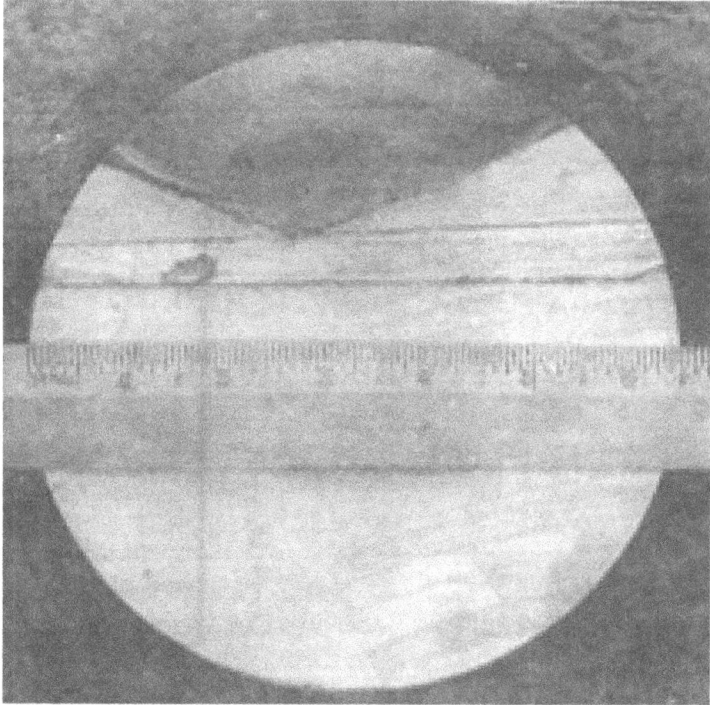

Fig. 7.10 Photograph illustrating a typical fatigue fracture in a large specimen.
(*Courtesy* : London Transport Board, Chiswick, U.K.)

1. Roughly speaking the fracture occurs suddenly, with no indication or warning.
2. Visually, the direction of fracture is at right angles to the direction of the tensile stress.
3. Very often, the fracture shows, up to a certain depth, a series of rings originating from the point of initiation of the failure. These rings are called the 'beach marks'. At this region, the fracture resembles the concoidal fracture generally observed in glass and minerals.

4. The fracture transforms into a ductile type in regions beyond those mentioned in (3). This shows that the part was unable to withstand the load once the crack has propagated to a considerable depth.

5. Above all, the point of origin of the fracture is clearly revealed. Generally the fracture starts at a surface defect or irregularity like a keyway, sharp corner, notch, root of a screw thread or an inclusion, which causes a stress concentration (Fig. 7.11).

Fig. 7.11 Fatigue cracks in a large machine gear. Note that the crack has initiated at the root of the gear.
(*Courtesy* : Magnaflux Corporation, Chicago, Illinois, U.S.A.)

Theories

There are many theories explaining the fatigue phenomenon. But not one of them is totally comprehensive and conclusive. However, failure due to the building up of the 'slip lines' at a particular region appears to be a reasonable proposition.

Orowan's Theory

According to Orowan, metals consist of minute 'weak' regions. The deformation in these regions is more than in the neighbouring 'strong' regions, under repeated loading. The strain increases as the value of the load increases. When the load or stress applied is such that it exceeds the value which the weak region can sustain, a crack develops.

Orowan's 'weak' regions may be those regions where the orientation of the grains is favourable for slip. They may also be the areas of high stress concentration like cracks, roots of the notches, screw threads.

Again, when a crack is developed it becomes another weak region. The process repeats itself resulting in the formation of fresh cracks. These get aligned suitably and result in a larger crack, leading to fracture.

Wood's Concept

When loaded, metals deform under conditions of slipping (Fig. 7.12a). When the load is reversed, the direction of slipping of the metal is also reversed (Fig. 7.12b). As a result there is every possibility for a notch to develop when the load reversal is repeated many times. This notch may be the root cause for the final fatigue failure.

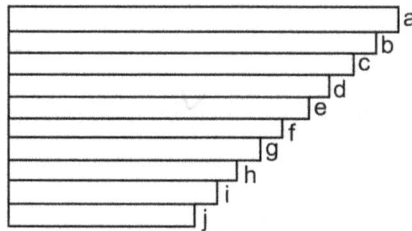

(a)

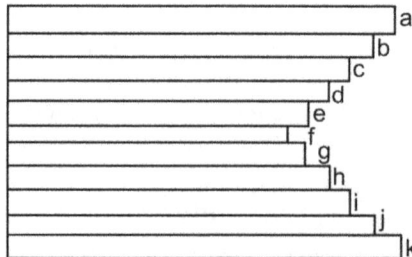

(b)

Fig. 7.12 Wood's concept.

This theory sounds reasonable because it establishes that – (1) the fatigue failure starts from the surface, and (2) the cracks initiate at the slip-bands, intrusions and extrusions.

EFFECT OF IMPORTANT VARIABLES

The following factors have a pronounced influence upon the fatigue property of metals:

1. Composition,   2. Stress concentration,
3. Size,   4. Surface condition and
5. Mechanical working.

1. Composition

The fatigue strength of a metal or alloy is not directly influenced by the alloying elements. However, those which increase the static (tensile) strength are found to increase the fatigue strength also. In the case of steels, carbon is found to have the maximum effect on the fatigue strength; carbon increases the fatigue strength.

It is further established that the alloys which undergo strain-ageing (work-hardening) exhibit a sharp knee in the S-N curve. In the case of steels, the sharpness of the knee is found to decrease and the curve flattens out as the carbon and nitrogen contents are decreased. Absence or decrease in the two alloying elements which are interstitial in character, means that the steel is less prone to work-hardening.

2. Stress Concentration

Fatigue strength is seriously affected by the presence of stress raisers in the specimen. Notches, holes, keyways, etc., which are found to initiate the fatigue failure, are all stress raisers. But all the engineering components invariably contain these stress-raisers. It is impossible to manufacture a bolt without threads, and without the threading, the bolt itself is useless. However, it is advisable to avoid those that are avoidable.

Besides, the changes in section and surface irregularities like machining cracks, porosity, inclusions, decarburised regions, etc., have their own influences in this regard.

A surface irregularity will act similar to a notch. The effect of a notch on the specimen is discussed in detail under impact testing. The stress concentration is increased at the root of the notch and thus a gradient of stresses is set up from that region towards the axis (centre) of the specimen.

The effect of a notch on the fatigue strength can be found out by testing sets of the similar specimens of the same material, one set notched and the other unnotched. The ratio between the endurance limit of the unnotched specimens and that of the notched specimens can be computed to give what is known as the 'fatigue notch factor K_f'. It is experimentally established that this factor varies with:

(i)   the severity of the notch,
(ii)   the shape of the notch,
(iii) the stress value,
(iv)   the type of the stress (or loading), and
(v)   the material.

3. Size

Unlike tensile strength, the fatigue strength is dependent upon a great many factors. Most important of them is the reversal of the stress. Though in all stress reversals the stress cycle changes from a tension peak to the compression peak, its effect is more severe on a thin section than in the case of a thick object. In the illustration (Fig. 7.13), value of d (the distance between surface and the axis) is less than D. Thus, the stress (or strain) gradient from the surface to the centre in the case of a thicker specimen is more. Also the volume of the material deformed is more. Further, increased size results in an increased surface area. Thus, there is more likelihood that a thicker section will fail.

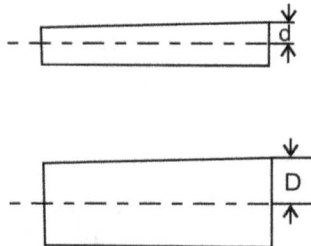

Fig. 7.13 Specimens of different sizes.

An exacting study of the effect of size on the fatigue is very difficult, for, it is practically impossible to prepare geometrically similar specimens of varying sizes having the same metallurgical characteristics like the structure and stress distribution. Loading thicker sections is another difficulty, which demands machines of greater capacities. However, the effects are studied within the limitations and it is established that the thicker the section, the lower is the fatigue strength. The fatigue strength is found to decrease by 25% when the thickness of the specimen is increased by 1/4″ above 1/2″. However, no marked decrease was noticed above the section size of 3/4″. It is further seen that in the case of steels, the size effect became more prominent as the carbon content is increased.

4. Nature of the Surface

Surface is the source wherefrom the fatigue failure originates. The condition of the surface has a pronounced effect on the fatigue strength of a metal.

It is established that different surface finishes produce different fatigue strengths for the same metal. The smoother the surface is, less is the scope for the presence of machining cracks or other surface irregularities. The surface roughness* is expressed in microns. As the surface roughness figure decreases the fatigue strength increases. In general, a buffed surface can be regarded as smooth, and an ordinary machined surface as rough. The fatigue strength of the same steel is observed to be 20-25 % less when tested in the ordinary machined condition than when the specimen is finished by buffing.

It is observed that the fatigue strength of a metal or alloy is remarkably increased if its surface is made harder. Logically it stands to reason, for it is difficult for a hard region to give way, and hence the scope for the initiation of the first fissure is minimized. Steels are case-hardened for improving their fatigue strength.

In general, electroplating is found to decrease the fatigue strength of steel. However, soft cadmium platings are made with no detriment to the fatigue strength. The effect is governed by many other factors in the platings, such as the nature of the plating, adhesion, residual stresses, induced stresses, etc.

---

* Separately treated under surface measurements.

5. Residual Stresses

Residual stresses are those stresses which still remain in the object after the process of manufacture is completed. These have very low magnitudes but at the same time manifest themselves very dangerously, for, they act on microscopic areas.

The origin of the residual stresses may be:

1. plastic deformation during which the material flow is not uniform throughout,

2. improper and insufficient annealing or normalizing after the plastic deformation,

3. quenching,

4. improper and insufficient tempering after quenching,

5. some finishing operations like machining, threading, boring, etc., and

6. sudden heating and cooling or excessive momentary loading, leading to improper recovery during the service.

It is said that the nature of the residual stress in a plastically deformed metal will be opposite to that by which the deformation was made. A metal which is plastically deformed in tension will consist of residual stresses of compressional nature and vice versa. For all practical purposes, the residual stress can be considered as identical to an externally applied stress. The maximum intensity of the residual stress in a metal is equal to its elastic limit.

A compressive residual stress pattern at the surface of an object increases its fatigue strength remarkably. This is the most effective method by which the fatigue strength of a metal can be improved. This is accomplished by:

(i)   shot peening,

(ii)   surface rolling using contoured rolls, and

(iii) in the case of a single object, the surface is just hammered lightly by a hand hammer.

It goes without saying that excessive peening or rolling will damage the fatigue strength instead of improving it. A series of laboratory tests and experience are warranted before a process is standardized for routine production work.

Another point that should be noted in this connection is that not all the aforesaid processes yield residual stresses of compressive nature. Those processes which give rise to tensile stresses should be avoided. If imminent, it should be followed by another treatment to improve the fatigue strength. Grinding can produce both the tensile and compressive types of stresses. Any process should be carefully controlled to avoid tensile type residual stresses.

INTERPRETATION OF FATIGUE TEST DATA

Fatigue test results obtained from the laboratory testing of polished small fatigue specimens cannot be taken as the sole basis for engineering design. It is opined that the fatigue test should be made on sample 'parts' preferably than on standard specimens. Further, it is suggested that the testing be performed, if possible, under the stress cycles similar to those the parts are expected to experience during their life. The number of stress cycles the part is tested for is arrived at by a prior estimate of the service life of the engineering component.

Again the test results should be considered only as a basis for design and not as statements of precision. Over and above the test data, comes the factor of safety. The factor of safety is fixed to take over the uncertainties like the corrosive media and the stresses the engineering member is likely to withstand during its life.

In laboratory tests on actual structurals and machine parts, loads of various magnitudes can be employed so that the range of stress during a stress cycle will approximate the stresses experienced by the part in actual usage. Thus, different special purpose fatigue testing machines are designed, differing in designs and capacities depending upon the purpose.

Tests of the aforesaid nature are both cumbersome and costly. That is why attention is again focused on the laboratory test data. Conventional fatigue test data can be relied upon for design purposes by suitably adapting it for the particular purpose. This adaptation includes making allowances for size, shape, surface condition, presence of stress raisers, corrosive atmosphere and the stress range in the actual service.

For the purpose of arriving at the due allowances, the criterion which is called the 'strength reduction factor' is determined. Reduction of the strength due to the variations in the shape or other things is thus duly taken care of. A series of fatigue tests are made on both notched and unnotched test specimens keeping the critical section the same as that of the engineering member. *S-N* curves are drawn for these tests. The ratio of the endurance limit for the unnotched specimen to that of the notched specimen is calculated, which gives the strength-reduction ratio. This is true for the particular size and shape of the notch and the given critical size of critical section.

Another design consideration is the 'stress concentration factor', a discussion about which falls out of the track in the present discussion. However, in general, it is found that the strength-reduction ratio is less than the theoretical stress concentration factor and so the former can be safely followed in design.

With all the above, the effect of size and surface condition should always be taken into account and due consideration shown for them.

Fatigue Failure in Service

Fatigue failures are most common in shafts – turbine shafts, main shafts of automobile and naval vessels, etc. The keyways, changes in section, screw threads and oil-holes are the general originators of the cracks and the subsequent fatigue failure. Very interesting fatigue failures occurred in the mid-fifties due to which two B.O.A.C. Comet aircraft were lost. The plane wreckages recovered from the sea beds were investigated into and it was determined that the wreckage resulted due to the fatigue failure of the fuselage. The stresses responsible were assumed to be those induced by the cabin pressurizing cycles for each flight. It was assumed that the cracks originated at the corners of the aerial windows.

CHAPTER 8

Solidification of Metals

The mechanism of solidification of metals and alloys is a good guide to producing the sound castings. Most of the foundry alloy characteristics can be studied by confining to binary system. Thus, the solidification proceeds in most of the alloys on either of the following processes,

1. At a constant temperature (e.g., Pure metals and eutectics)
2. Over a temperature range (e.g., solid solutions)
3. By a combination of the solidification over a range of temperature, followed by a constant temperature freezing.

The foundry metal is poured at a higher temperature than its melting point. This amount of additional temperature (over the melting point) is referred to as the superheat and is essential to facilitate handling. The metal thus, has to undergone three phases of cooling during solidification as follows:

(a) Cool from the temperature of pouring to its solidification.
(b) Give out its latent heat of solidification while remaining at the melting point over melting temperature range; and
(c) Cool from the solidification temperature to the room temperature.

During all the above three phases the metal undergoes a contraction. The total shrinkage is the sum of the above three.

Freezing of a pure metal: When a pure metal is filled in the mould cavity, the portion of the metal which first reaches the freezing temperature starts to solidify. Generally this portion will be that near the mould surface. The chilling of the mould surface results in the formation of a thin skin or shell of the solid metal. As the heat abstraction continues, the liquid metal continues to freeze on to it and the thickness of the shell increases. The growth of the solid metal proceeds towards the centre of the casting, depending upon the prevailing temperature gradient. The interface between the solid and liquid is relatively smooth, because the metal is freezing at a constant temperature. It is found that the thickness of the skin solidified at any time can be expressed by the empiricial relation ship,

$$D = K\sqrt{t} - C$$

where

K = a constant determined by the size of the casting,

C = a constant depending upon the degree of the superheat

and   t = time in seconds.

Dendritic Growth: Growth by the formation of dendrities is very common in metals. A needle or a spine of solid metal grows first. As this

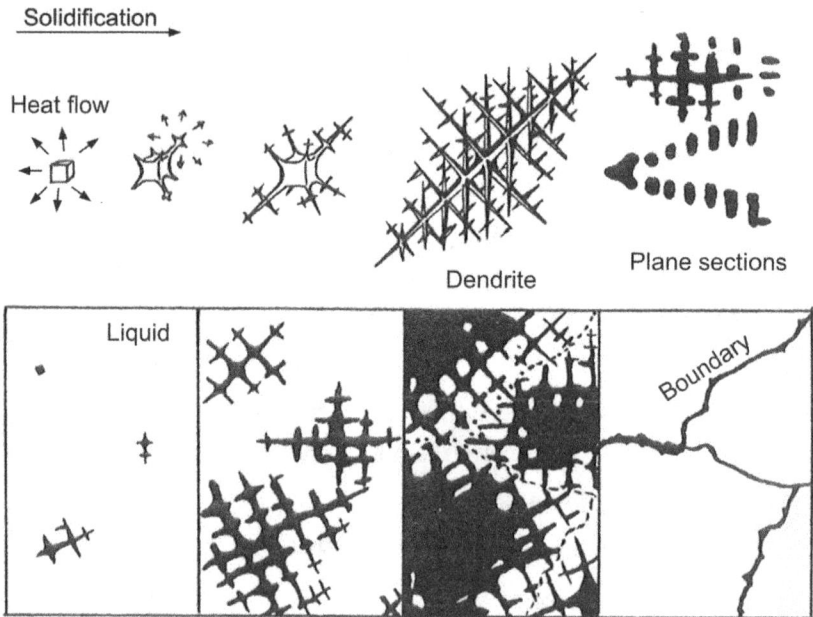

Fig. 8.1 Solidification of a metal. For f.c.c and b.c. metals dendrite arms extend in the cube face [100] direction.

needle enlarges, additional needles grow from it, at right angles. On these branches, further grow other needles and as a result a skeleton tree like structure can be seen and this is called a dendrite. The size and shape of the dendrite is determined by its surrounding dendrites. Each dendrite can be taken as the skeleton of a grain. In the chilled crystal layer referred to above, a large number of dendrites are nucleated and grow in all directions. They come into contact with each other. Thus their multi-directional growth is hindered. Hence the fine equi-axial structure.

The second stage of freezing consists of the growth of the solid/liquid interface into the liquid. The dendrites which are suitably situated in the outer skin, grow into the liquid (at right angles to the surface). The result is that the dendrites become elongated or appear columnar.

The solid or liquid interface now appears rough.

The chilling effect and directional heat abstraction will no longer prevail after some time.The solid already formed and the central residual liquid will be in equilibrium. The solidification occurs by nucleation and growth with the formation of equi-axial grains at the centre. This is the third stage of solidification.

Shrinkage: (a) The liquid-liquid shrinkage (b) the solidification Shrinkage (c) and the contraction in the solid state; are illustrated in the Fig. 8.1(a).

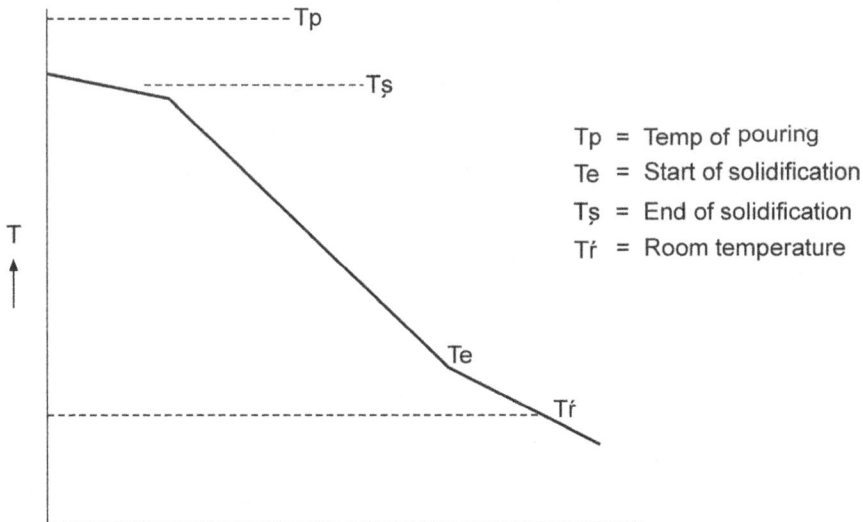

Tp = Temp of pouring
Te = Start of solidification
Tş = End of solidification
Tf = Room temperature

Fig. 8.1(a)

Mechanism of Solidification: Solidification or crystallization is the change of state from liquid to solid. It occurs at a constant temperature in metal and eutectic alloys. In the case of alloys which are not eutectics, solidification takes place over a lower range of temperature. However the solidification process comprises of two stages.

1. Nucleation
2. Growth

Nucleation: As the temperature of the liquid decreases, the atom movement also decreases. In other words, the atom vibrates over a short distance about a mean position. These distance reduce to that of solid state and a micro-solid forms. As the above process proceeds both potential and kinetic energies, decrease with the decrease in temperature. It is a well known fact that the stable state of a system is that of its lowest energy, micro-solids start to form in the liquid. These micro solids are called nuclei and the process, nucleation. The nuclei are unstable and will be breaking up, melt in the liquid. At the same time, more nuclei will also form. After a very short time, many more nuclei will also form. Afterwards formation of nuclei predominates their destruction. Nuclei are first solids formed in a liquid.

Growth: Each nuclus starts to become bigger due to the gradual deposition of solid into it. This occurs in all the three directions, but it will be (may be) more pronounced in one direction. These are the dendrites referred to above and the increase in their size is called growth. Growth of the dendrites continues until the neighbouring dendrites, growing themselves, come to touch them and stop future growth, some remaining liquid will ultimately solidify in the gaps between the grow dendrities, called grains. The last solidifying solid does not have any set pattern of atomic arrangement, and will not be crystalline. These grain boundaries are thus the regions of haphazard atomic arrangement.

Solidification of an alloy: In actual industrial conditions equilibrium cooling is extremely slow cooling is seldom practical nor practiced. Equilibrium cooling rate can be taken to be a cooling rate of 1°C per minute. When the cooling of an alloy under faster cooling is non-equilibrium cooling conditions is studied, the solidification pattern will be different. Let us study the solidification of a solid solution alloy of 25% B in an alloy system AB. Refer to the (Fig. 8.2) solidification of the alloy from the liquid state starts at T_1 with the formation of a solid of composition S_1. When the temperature reaches T_2, the liquid composition will be L_2 on the liquidus line while the solid forming will have a composition of S_2. And at a still lower temperature T_3, the composition of the liquid will be L_3 and that of the solidified solid will be S_3. The process continues as the temperature drops, the composition of the solid separating will depart still further from the equilibrium value. In non-equilibrium cooling, the solidus line will assume an imaginary shape, dipping more than the equilibrium curve shown in the dotted line in the figure. The solid separating will have a composition of S_3^1 and not S_3 as shown by the equilibrium diagram. But the liquid composition will follow strictly the liquidus line because the atomic movement will be faster in the liquid state. The solidification completes at T_3^1 instead of T_3, due to the super cooling mentioned above, the solid separating i.e., the last to identify solid will be having a composition B^1. Thus, when we microscopically examine the chemical composition of grain from the centre to the surface there will be a change in composition from S_1 to S_3 layer by layer, each layer being richer in the percentage of A than it should be ideally. This variation in composition of an alloy from core to the surface due to non-equilibrium cooling is called coring.

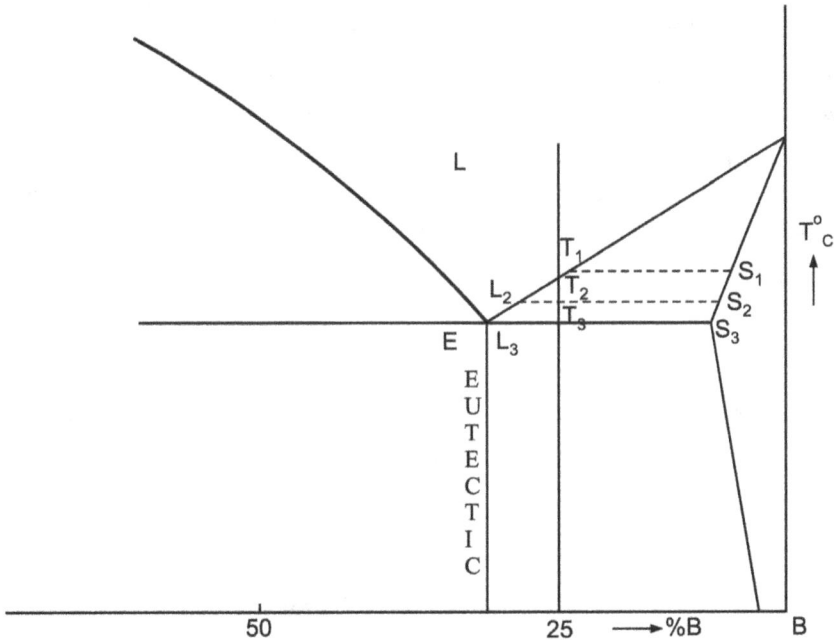

Fig. 8.2

CHAPTER 9

Solid Solutions

Metals form solutions with each other. They also form compounds at different compositions. In the case of metals forming solutions, the word 'solution' appears strange. The metal solutions are called solid solutions, to differentiate them from the commonly known liquid solutions of sugar or salt in water. As far as physical properties are concerned they are similar. Just as we can not differentiate between sugar solution and water with naked eye, metal solutions are also not differentiated easily at room temperature.

Compounds between metals is also common when two metals do not possess favorable atomic properties to form solid solutions, they form chemical compounds. These are called intermetallic compounds. They can be studied by studying their physical properties.

Factors determining solid solubility: Solid solutions are classified into two types, namely substitutional and interstitial.

Substitutional Solid solutions: In these solid solutions, the solute atoms occupy the lattice sites of the solvent atoms. Solute atoms substitute the solvent atoms at some lattice sites. Hence the name.

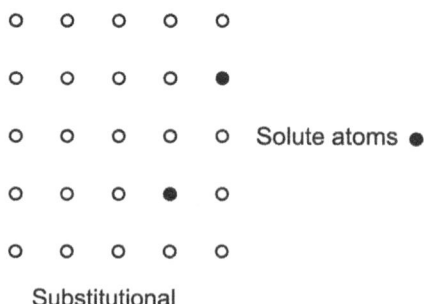

Substitutional

Fig. 9.1 Substitutional solid solutions.

Interstitial Solid Solutions: In these solid solutions, the solute atoms occupy places in between the solvent atoms in the lattice. In the figures

9.1 and 9.2, atoms denoted by the letter A are solvent atoms and by letter B, solute atoms.

Solid solutions

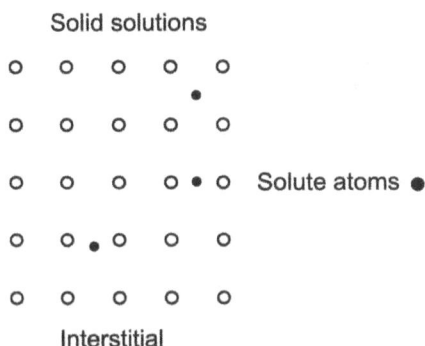

Fig. 9.2 Interstitial solid solution.

Whether two metals A and B form a substitutional solid solution or an interstitial solid solutions, can be predicted by Hume Rothery rules, as they are commonly known as. They are four propositions, as detailed below.

1. **Crystal Structure factor:** Complete solid solubility of two elements is not possible unless both possess the same type of crystal lattice structure. Two metals A and B will form solid solutions only if both have FCC structure or both have BCC structure like that.

2. **Atomic size factor:** For metals A and B to form a substitutional solid solution, their atomic sizes should not differ by more than 14%. If the size difference is less thatn 14% but more than 8%, the system usually shows a minimum percentage solid solubility.

3. **Chemical affinity factor:** The greater the chemical affinity of the two elements (metals) less is the possibility of solid solution formation. On the other hand they tend to form chemical compounds. Generally, the chemical affinity can be studied from the periodic table. Elements which are further apart in the periodic table, possess more chemical affinity.

4. **Relative valency factor:** If the solute metal has a different valency than the solvent metal, the number of valence electrons per atoms known as electron ratio will be changed. A metal of lower valency dissolves a metal of higher valency more than vice versa. If we consider Aluminum - Nickel system the valency of nickel is less

than valency of aluminum and so dissolves up to 5% aluminum. Aluminum which has a higher valency can dissolve nickel only upto 0.04%.

It can be understood from the above that the lattice structure of the solid solution will be that of the solvent metal. Also there will be a slight expansion taking place if the solute atom is larger than the solvent atom. Like wise contraction occurs if the solute atom is smaller than the solvent atom.

Interstitial Solid solutions: These are formed when the atoms of smaller atomic radius fit into the space in the lattice of a solvent whose atoms are relatively big. As you are aware of the spaces existing as interstices in atomic structures is very restricted. Only atoms of very small atomic radius form interstitial solid solutions. Hydrogen, Boron, Carbon, Nitrogen and oxygen atoms whose atomic radius is less than $1°A$ from interstitial solid solutions.

This type of solid solution differs from an inter metallic compound. An interstitial solid solution will have only a small solute percentage. Even this small percentage also is variable. An interstitial compound will have more of the 'solvent' and has a fixed composition. The solute atoms in the lattice will have more mobility due to their small size and more in the interstitial spaces of the lattice. Lattice structure always shows an expansion. These solid solutions are rare; of course the solution of carbon in iron is an exception. FCC form of iron ($\propto$ iron) dissolves up to 2.0% carbon at 1150°C while the solubility of carbon in BCC iron (δ) is only 0.025% at 723°C.

It can be seen from Fig. 9.2 that there will be distortion of the lattice in the region of the solute atoms. This distortion will interfere with the movement of dislocations on the slip planes. Thus, the alloy is stronger.

Interstitial Compounds: Of the three of compounds formed and elements, our attention will be readily drawn to the interstitial compound because there will be a metal as one of the constituents in these compounds E.g., TiC, Fe_4N, Fe_3C etc. It should be noted that the metals involved this way are transitional metals. These compounds are high melting and extremely hard, besides being metallic.

Electronic Compounds: When we study the equilibrium diagrams of copper, gold, silver, iron and nickel with calcium, magnesium, zinc and aluminum, some interesting similarities can be found. A number of intermediate phases can be seen in these diagrams with similar lattice structure. Hume Rothery had suggested that these intermediate phases occur at or at about the same compositions in each system that have a fixed ratio of valence electrons to atoms and called electron compounds. The table below gives some examples.

Electron Compounds

BCC LATTICE Electron/atom 3/2	COMPLEX CUBIC Electron/Atom 21/13	H.C.P Electron /Atom 7/4
AgCd	Ag_5Cd_8	$AgCd_3$
AgZn	Cu_9Al_4	Ag_5Al_3
AuMg	$Cu_{31}Sn_8$	$AuZn_3$
Cu_3Al	Au_3Zn_8	Cu_3Si
Cu_5Zn	Fe_5Zn_2	Fe Zn_7
Fe Al	Ni_5Zn_{21}	Ag_3Sn

Consider the compound AgZn, silver atom has one valence electron while the zinc atom has two. Thus two atoms of the compound have three valance electrons i.e the electron atom ratio is 3/2.

While making such calculation, it should be assumed that iron and nickel have zero valence electrons.

The existence of electron compounds is a consequence of the nature of the electronic bond. The stability of such a phase mainly depends upon the electron concentration and the pattern of atomic sites in the crystal structure while the actual distribution of different atoms in these sites is not so important.

CONSTITUTION OF ALLOYS

Material properties are classified into two types,

    1.   Structure insensitive properties

and 2.   Structure sensitive properties

Structure insensitive properties are electric and thermal conductivities, elasticity etc. The microstructure of the material does not effect the properties in anyway and hence the name. On the other hand, properties like strength, hardness and the ductility which are of importance to the serviceability of the material are structure sensitive. The microstructure of the material actually controls these strength properties. Here by structure we mean the microstructure of the material.

Metals and alloys contain grains and grain boundaries in their microstructure. The composition, occurance, size and shape of the grains and their orientation - all comprise of the term microstructure needless to say that the grain structure will be changed by changing the composition of the alloy. The phases, their size and orientations can be changed by judicious variation of cooling and heating cycles. This is called heat-treatment. All this information is studied and stored. It is found that the best way to do this is to plot the phase or equilibrium diagrams.

Before proceeding further, we should understand and get an idea about the equilibrium of a system and how it can be defined. Three important variables define the equilibrium of a system. They are pressure, temperature and composition. Gibbs has candidly formulated the phase rule which shows these variables are related, thus,

$$P + F = C + 2$$

where   P = Phases

C = Components

and   F = Degrees of Freedom

The application of phase rule to systems of heterogeneous equilibrium is illustrated in Fig. 9.3.

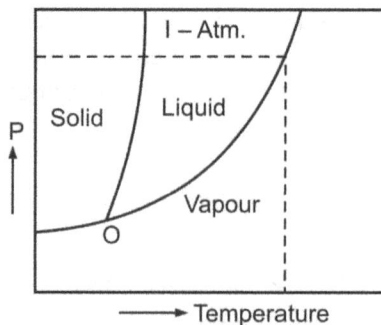

Fig. 9.3 Ice water steam system.

Let us consider the solid, liquid and vapour equilibrium of a one component system. Here the single component taken is water. For a single phase, water, $C = 1$ and $P = 1$. So, the degrees of freedom $F = 2$. This is a bi-variant equilibrium (Fig. 9.3). It means that the temperature and pressure can be set or changed arbitrarily. For water-water vapour, water-ice or ice-water vapour equilibria,

$C = 1$;

$P = 2$; and $F = 1$

That is, they are univariant, meaning that pressure or temperature is the independent variable, when all the three phases ice, water and water vapour are in equilibrium, $F = 1$. This is an invariant system. Using this information, a pressure temperature phase diagram for a single component system drawn schematically is shown in Fig. 9.3 curves separating two single phase regions correspond to two phase univariant equilibria. At the triple point 0 (trivariant) three univariant curves intersect.

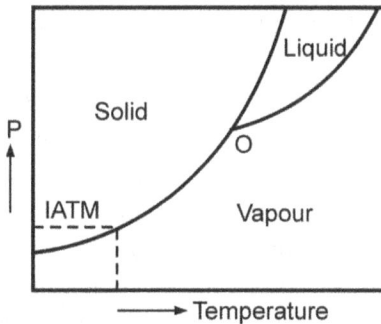

Fig. 9.4 Solid-liquid-gas CO_2 system.

If the pressure for the univariant system is greater than 1 atm (See Fig. 9.4), the substance does not melt, solid CO_2 sublimes at -78.5 °C at atmospheric pressure. If the pressure is increased to 5.11 atm, it melts at -56.4 °C. Again it follows that at a constant pressure in a single component system, the transformation of the substance from one crystalline form to the other (melting or boiling) occurs at particular temperatures, characteristic of the substance. Further it should be noted that the temperature of the substance remains constant throughout the transformation.

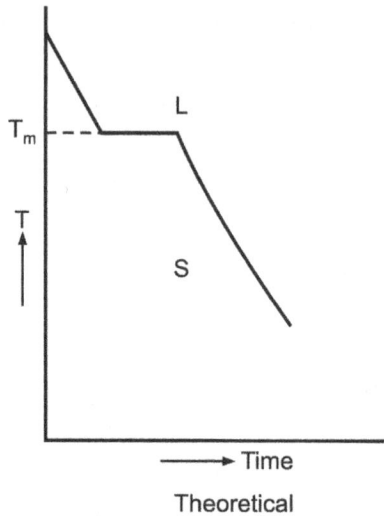

Fig. 9.5 Cooling curves.

The above phenomenon is shown in the temperature time cooling curve (Fig. 9.5). The horizontal line corresponds to the solidification temperature of the single component system at a constant pressure. Increasing pressure increases the temperature of the phase transformation as shown in Fig. 9.4. However, this effect is small on solid-solid transformations and melting temperatures. Diffusion and nucleation play a dominant part in the process of transformations. In real systems, a certain amount of undercooling occurs before solidification sets in the melt is maintained in a metastable state at a temperature below the freezing point and most of the cooling curves are as shown in the Fig. 9.6. This is due to the letting out of the latent heat. Its extent is however dependent upon several factors besides the latent heat.

Phase Equilibrium Composition diagram: In order to completely specify the state of a system in equilibrium, it is necessary to specify three independent variables. These variables, which are externally controllable are the temperature, pressure and composition. With pressure taken as I atm, the metal equilibrium can be expressed in terms of two independent variables viz., temperature and composition. A diagram plotted for an alloy system in this fashion is called an equilibrium diagram or phase diagram. Thus, equilibrium diagram may be defined as a plot of the composition of phases as a function of temperature in any alloy system under equilibrium conditions.

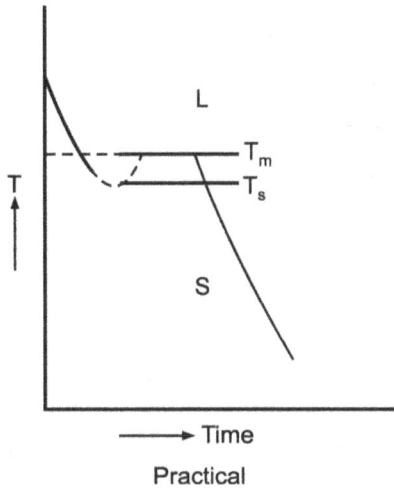

Fig. 9.6 Cooling curves.

Co-ordinates of the phase diagrams: Temperature usually forms the ordinate for the phase diagrams. The alloy composition in weight percentage forms the abscissa. It is sometimes convenient to express the composition in atomic percents. Conversion from weight percentage to atomic percentage can be made from the following relationship,

$$\text{At percentage of} \quad A = \frac{100\,x}{\left[X + Y\left(\dfrac{M}{N}\right)\right]}$$

$$B = \frac{100\,x\left(\dfrac{M}{N}\right)}{\left[X + Y\left(\dfrac{M}{N}\right)\right]}$$

where, M = Atomic weight of metal A.

   N = Atomic weight of metal B.

   A = Wt percentage of metal A.

   B = Wt percentage of metal B.

EQUILIBRIUM DIAGRAMS

There are different methods of plotting equilibrium diagrams. They are,

(a) Thermal Analysis

(b) Metallography and

(c) X-ray diffraction

Of the above, the simplest and most widely used procedure is thermal analysis. The process consists of plotting the cooling curves of different compositions in the system. For example, if an equilibrium diagram of the binary system between A and B should be studied, as many alloy compositions in the 0 – 100% B range as possible are taken and thermal analysis is made, care should be taken to cover the composition range in a representative fashion viz alloys A. 10% B, A. 25% B, A. 40% B, A. 65% B, A. 80%B, besides pure metals A and B are taken. Thus we get five alloys to study the equilibrium cooling patterns, besides A and B. Five, may be regarded as a minimum number to achieve a good spread over (representation) in the alloy range.

The alloy for which the cooling curve is to be plotted is melted in a suitable container. A temperature recording set up, say a pyrometer is immersed in the molten alloy. The temperature of the alloy is recorded at regular intervals, say every minute or two, until the room temperature is reached. Times and temperatures are tabulated and kept aside. Looking at the table with the alloy composition, the data will be recorded as in the table shown as a model. The experiment is repeated with an alloy of another composition x_1 and soon until all the selected compositions are cooled and the (Time) minutes – (Temp) °C time-temperature values are tabulated.

The next step in the thermal analysis consists of plotting cooling curves for each of the compositions analysised thermally. These cooling curves are plotted on the same graph sheet with a common X-axis and two Y-axes, representing the two constituents of the alloy, A and B.

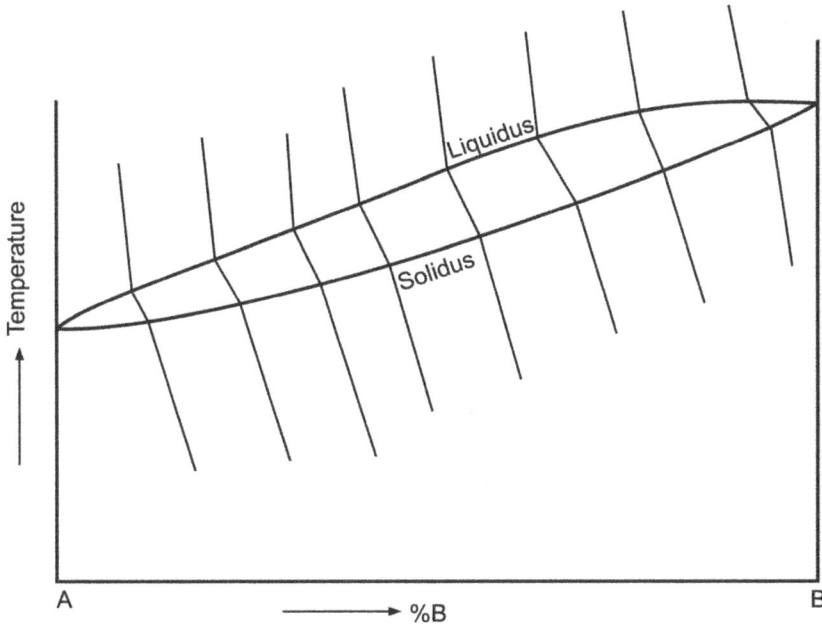

Fig. 9.7 Thermal analysis.

It is imagined for an easy understanding that the two metals A and B are completely soluble both in liquid and solid states. We get a series of cooling curves as shown in Fig. 9.7. When we join all the points where solidification of the alloys had started by a smooth curve and also the points where solidification was completed by another curve to the melting points of A and B, we get a figure resembling a convex lens. This is the equilibrium diagram of AB.

The upper line is the line obtained by joining the various points of the start of solidification is called the 'liquidus' and the lower line where the points of ending of solidification be is called the 'solidus'. The equbrium diagram consists of two points, two lines and three areas. The two points are the melting points of the metals A and B, the two lines are solidus and liquidus and the three areas are the area above liquidus representing liquid, between liquidus and solidus representing solidifying zone consisting of both liquid and solid and the area below the solidus line, representing the solid phases.

Determination of chemical composition of the phases, Lever Rule: It is common knowledge that a 25% B alloy means it contains A-75% and B-25% in its composition. But, owing to the separation of solid from the liquid and the process going on for a period over the liquid-solid temperature range, it will be interesting to study the relative quantities and their chemical compositions (of the solid and liquid phases) during the solidification stage.

Lever rule as it is commonly known, comes in handy to calculate the compositions of the two phases at any specified temperature. A temperature horizontal is drawn at the given temperature to interest the solidus and liquidus (see Fig. 9.8) and these points are dropped on the x-axis to read the composition of solid separating and that of the remaining liquid.

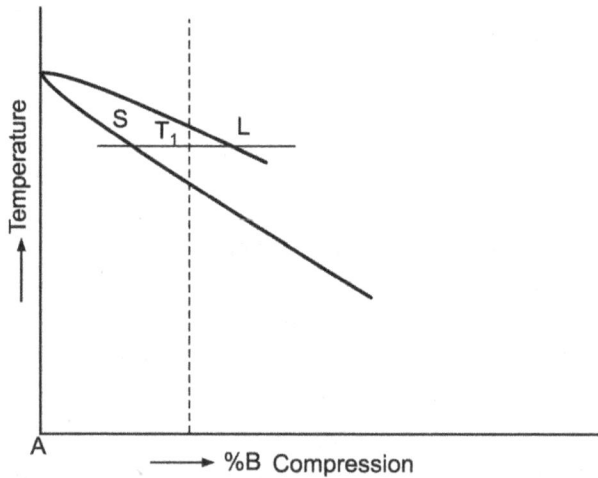

Fig. 9.8 Lever rule.

Not only the two compositions, but their relative proportions can be calculated by applying what is commonly known as the lever rule. In this example, the proportion of solid separated is T_1L/L_5 and the residued liquid is ST_1/L_5. The tie line acts as a lever and the temperature point as it acts as a fulcrum in this computation and hence the name Lever Rule.

We have taken an example of a system between A and B in which the metals A and B are completely soluble in each other both in liquid and solid states, for the sake of simplicity and easy understanding of plotting an equilibrium diagram. In actual practice, the equilibrium systems are more and more different and some complicated. Without going into the complexities of binary systems i.e. systems containing two metals, we can chronicle some important and basic binary systems as follows.

1. Two metals are completely soluble in liquid state as well as in the solid state.

2. Two metals completely soluble in liquid state but insoluble in solid state with a eutectic.

3. Two metals completely soluble in liquid state but partially soluble in solid state with a eutectic.

4. Two metals completely soluble in liquid state and with the formation of a congruent melting intermediate phase.

5. Two metals completely soluble in liquid state and decomposing into two solids by peritetic reaction.

6.   Two metals partly soluble in liquid state (monotectic reaction).

7. Two metals insoluble in both the liquid and solid states.

8. Solid state transformations like allotropy, order-disorder, eutectoid reaction and the peritectoid reaction.

1. *Equilibrium cooling of a solid solution alloy:* The alloy is similar to the one we have considered to study the plotting of an equilibrium diagram. We will consider the mode of solidification of a particular composition in the system; say A, 25% B. At T, the alloy is a homogeneous single phase of two metals, dissolved in each other. It remains like that until the temperature TL is a print on the liquidus line and solidification starts at that temperature. The first formed solid will have a composition, richer in the higher melting point metal B. Progressive cooling makes the liquid richer in lower melting metal A, its composition varying along the liquidus. When the temperature reaches to the FG whole alloy has solidified. It consists of grains and grain boundaries. Eutectics, on microexamination reveal this structure of grain and grain boundaries, like pure metals.

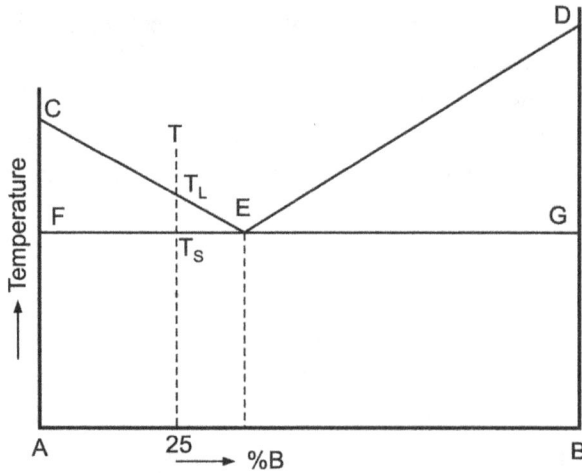

Fig. 9.9 Eutectic system.

A variation of the above explained eutectics is called pseudo eutectics. The solidus and liquidus lines go through a minimum or maximum point, an alloy of such a composition behaves just like a eutectic or pure metal (Fig. 9.9). It starts solidification and also ends solidifying at a constant temperature with the same composition. Alloys which show a minima x are Cu-Au and Ni-Pd, but so far there are no practical examples of occurring as maxima in any alloy system; as a maxima in the system is hypothetic only.

2. Two metals completely soluble in liquid state and completely insoluble in solid state. Raoult's law states that the melting or solidifying point of a substance is lowered by adding a second substance provided the second substance is soluble in the first one when liquid and insoluble when solidified . The amount of lowering of the solidifying temperature is proportional to the molecular weight of the solute.

Let us consider two metals A and B. As per Raoult's law when a quantity of B is added to A, its melting point is lowered. Just as we regarded B being added to A, we can take into account that A is being added to B, reducing the melting point of B. Thus two liquidus curves come into existence. The point where these two meet is the alloy composition and has the minimum freezing point in the system AB. This point is known as the eutectic and the composition it symbolizes eutectic of composition. If we draw a horizontal line at the eutectic point touching the two Y axes, it becomes the solidus line. In the Fig. 9.9 CE and ED are the liquidus lines and FE and EG are the solidus lines. E is the eutectics. The alloys existing to the left of E are known as hypo-eutectic alloys and those existing on its right are called hypereutectic alloys. The alloy of eutectic composition E solidifies into a mixture of two solids AB at a constant temperature. This is called eutectic reaction.

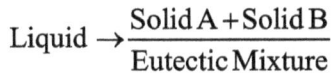

$$\text{Liquid} \rightarrow \frac{\text{Solid A} + \text{Solid B}}{\text{Eutectic Mixture}}$$

The phase diagram consists of four areas. The area above the liquidus line is a homogenous liquid solution. The remaining three areas are two phase areas. CEF consists of solid A and yet to solidify liquid. Similarly, the area DEG consists of solid B and yet to solidify liquid. The field AB GE comprises of eutectic and the corresponding metal depending upon it is on the A side or B side of the eutectic. These ideas can be applied to any phase diagram while studying.

The eutectic mixture AB solidifies in alternate steps (micro level) of A and B, suppose A solidifies, the liquid becomes richer in B so, the next to solidify will be B. After a few moments there may be depletion of B in the liquid from the equilibrium composition leading to the solidification of A. This continues until the whole of the liquid has solidified. Thus, the eutectic mixture occurs (solidifying at a constant temperature). The eutectic solidification is incongruent because there will be a difference in composition between the liquid and the two solid phases.

3. Two metals completely soluble in the liquid state but only partially soluble in solid state. Most metals show some solid solubility for each other, this type of equilibrium is most common and important, (see Fig. 9.9). Alloys in this system solidify as α solid solution (not pure metal A) and β solid solution (not as pure B). In the equilibrium diagram, these exist at the ends. So, they are called

terminal solid solutions. In place of A there will be α and in place of B there will be β in comparison to the previous equilibrium, due to the formation of solid solutions. Let us eloberate the equilibrium solidification of a hypo-eutectic alloy and one hyper eutectic alloy, in addition to the eutectic composition (Fig. 9.9).

Solidification of Alloy (1): The alloy, (see Fig. 9.9) is liquid at T. The liquid cools upto TI, the temperature of the liquidus. At this temperature solid solution of the composition s starts to solidify. On cooling through the temperature range TI Ts, this process of formation of X continues with the composition of residual liquid varying along the liquidus line AE. At Ts, the solidification of X is complete. The temperature of the alloy which is primary solid solution and decreases until the temperature Tsl is reached. Further cooling, results in the x solid solution becoming richer in A, giving out eutectic mixture. This line along which the composition of the solid solution varies in solid state, is called Solveus Line. Thus at room temperature, the alloy consists of x and eutectic β.

Solidification of alloy (2): The alloy is liquid at T. The liquid continues to cool up to the temperature of the eutectic E is reached. At E, it solidifies as a eutectic, comprising of a solid solution of the composition A%B and β solid solution of A %B.

Solidification of alloy (3): The alloy is liquid at T. The liquid cools up to T_1. At this temperature, solid solution β of composition S_1 starts to solidify. On further cooling over the temperature range T_1 Te, β continues to separate along the solidus line DF while the liquid composition reaches the eutectic composition. Below Te all the liquid remaining solidifies as α β eutectic. On further cooling to room temperature, the solid solution β rejects some α along the solvers line. Hence the primary β at room temperature consists of %B only besides α β eutectic.

Mechanical Properties of eutectic alloys: In all metallic alloy systems the properties of an alloy are dependent upon its microstructure. Microstructure means the phases existing, their size, orientation and of course strengths. In alloy (1) and (3) described above, there occurs terminal solid solution in the structure either α or β besides the eutectic mixture αβ.

All the solid solutions are soft, ductile and possess lesser strength. But as the alloy compositions move towards the eutectic, the strength dominates and elongation decrease. Further the properties of an alloy in the system approximate to the properties of the phase which exists as a matrix in the microstructure. It is natural that the matrix or continuously existing phase will be that which makes up the greater proportion in the eutectic mixture, cooling rate of the alloy also plays a dominant role in controlling the properties. Faster cooling leads to the formation of a finer eutectic mixture: a fine structure possesses more strength (Fig. 9.10).

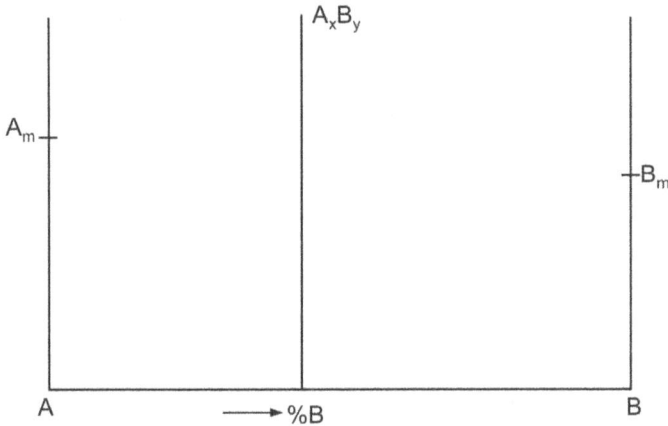

Fig. 9.10 Congruent melting phase.

4. **Congruent melting intermediate phase:** Congruent transformation or phase change is defined as the change undergone by one phase at a constant temperature and chemical composition, into a chemical compound. As you all know, pure metals solidify congruently(Fig. 9.10). Intermediary phases are named congruent phases because they occur between terminal phases in the equilibrium diagram. They are represented in the equilibrium diagram as vertical lines indicating the chemical composition – AxBy x and y are the subscripts showing the number of atoms in combination in the compound molecule (Fig. 9.10).

From the Fig. 9.10 it can be seen that the diagram is made up of two parts horizontally. One part is A and Ax By while the second part is AxBy and B. Each of these portion may be of any type studied so far. For example if both the portions show eutectic mixtures, the equilibrium diagram AB can be drawn as shown in the Fig. 9.10. The eutectic reactions may be written as,

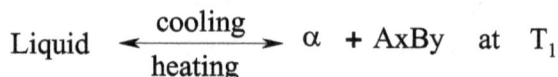

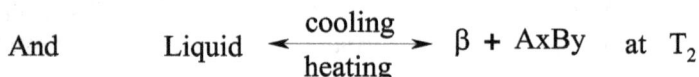

$$\text{Liquid} \; \underset{\text{heating}}{\overset{\text{cooling}}{\rightleftarrows}} \; \alpha \; + \; AxBy \quad \text{at} \quad T_1$$

$$\text{And} \qquad \text{Liquid} \; \underset{\text{heating}}{\overset{\text{cooling}}{\rightleftarrows}} \; \beta \; + \; AxBy \quad \text{at} \quad T_2$$

Bifurcation of the equilibrium diagram as above simplifies the study of the systems where congruent melting phases occur.

5. **Peritectic Reaction:** A liquid of a particular composition reacts with a solid of another particular composition to form another solid of a different composition isothermally.

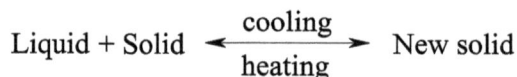

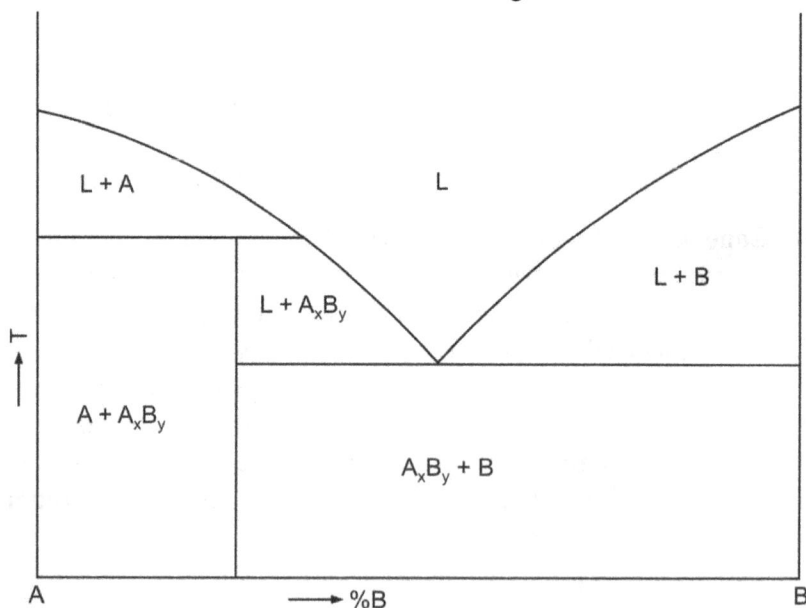

$$\text{Liquid} + \text{Solid} \; \underset{\text{heating}}{\overset{\text{cooling}}{\rightleftarrows}} \; \text{New solid}$$

Fig. 9.11(a) Peritectic reaction.

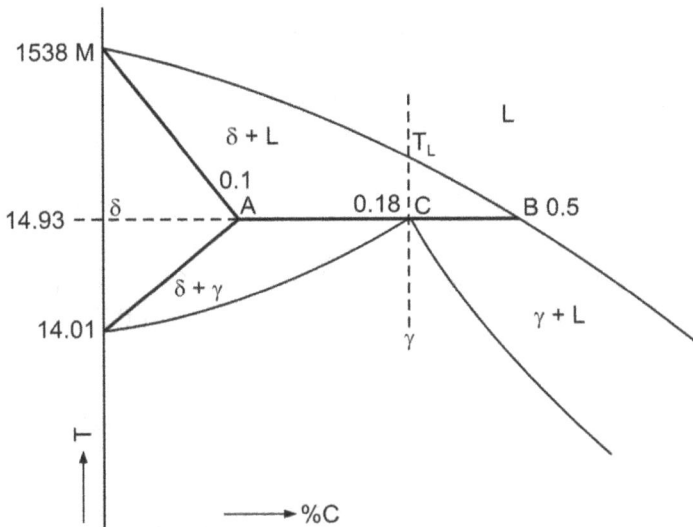

Fig. 9.11(b) Peritectic reaction in iron carbon system.

The new solid is usually intermediate phase. In some systems, it can be a terminal solid solution (e.g., Ag. Pt). In fact, the peritectic reaction may conveniently be regarded as a reverse of eutectic reaction (see Fig. 9.9). The peritectic reaction occurring in iron carbon system is a very common example (Fig. 9.11).

Of the many metals and alloys in use, steels occupy a unique position. Their main component iron, is abundantly available in the earth's crust and so available cheaply in large tonnages. Metallurgically speaking, the solubility of carbon in iron and the different products formed present a vast scope for study. This is due to differing solubilities of carbon in the different allotropes of iron make steels amenable to heat treatment. Consequently, heat treatment of steels has itself become a subject of study. More about heat treatment processes is given in succeeding pages. Peritectic reaction yields austenite which is the starting phase in many heat treatment processes. Further, austenite itself is made to exist at room temperature (austenitic steels). Such steels of different compositions are widely used in many industrial and domestic appliances.

Formation of austenite: Let us study the solidification pattern of a steel of peritectic composition itself. Liquid steel starts to solidify at TL (see Fig. 9.11) with the formation of δ-ferrite. On further cooling, δ-_ferrite composition varies along the solidus line MB while the composition of the balance liquid changes along the liquidius MO. When the temperature reaches NO, a state of equilibrium is reached. Equilibrium exists between δ ferrite of N% carbon and liquid of B% carbon. They combine and give rise to a solid of composition C, austenite. This is the peritectic reaction.

6. **Monotectic reaction:** So far we have discussed various equilibria wherein different phases occurring below the temperatures of the start of solidification. Also, it was assumed that the two metals are completely soluble in liquid state. However, there is a possibility of the two metals, not completely soluble in each other (immiscible) like oil and water. Thus, there exists a micibility gap in the liquid state in the equilibrium diagram.

A typical equilibrium of this type is shown in Fig. 9.12. In this diagram, the liquidus line is Ta Te JTb. Alloys consisting of compositions between C and F at temperature just above Tm will consists of two liquid solutions L_1 and L_2 is given by F. Here, we should understand how L_2 has formed. Owing to the continuous separation of A (while cooling up to the temeperature C) the percentage of B in the liquid L_1 increase, reaching to that level of L_2.

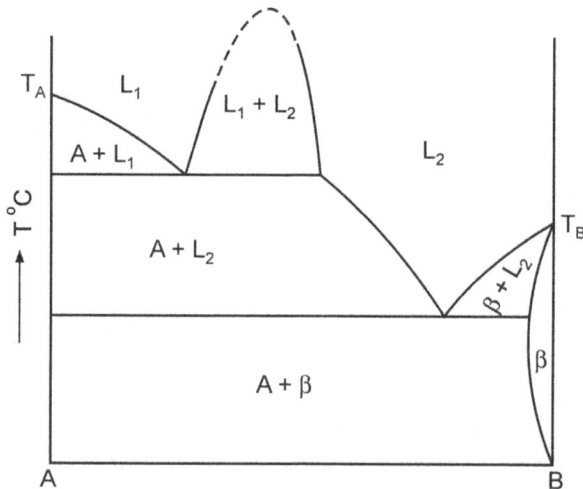

Fig. 9.12 Monotectic.

Still further cooling leads to the change in the composition of L_2 along FE. All the while, A continues to separate. At E, a eutectic mixture forms as A and β. Point C is similar to eutectic but instead of a eutectic mixture of two solids, a mixture of a solid and a liquid is obtained here. Hence the name monotectic.

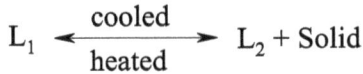

$$L_1 \xrightleftharpoons[\text{heated}]{\text{cooled}} L_2 + \text{Solid}$$

Copper lead equilibrium illustrates a monotectic (see Fig. 9.12). In this, the L_1 and L_2 lines exist as a closed line. The terminal solids are labeled as α and β. The mutual solubilities are so small that almost pure metals like lead and copper exist.

7. **Two metals insoluble in both liquid and solid states:** Many combinations of metals are there which fall into this category. Where in liquid state two metals exist in a liquid state albeit any solubility. It will be difficult to easily detect this. But when they are solidified i.e., the temperature of the liquid mixture is lowered to a temperature below the lower the melting metal. The solidified solid mass possessing two distinct layers with a well-defined (mutual) contact surface. This type of equilibrium systems is shown in Fig. 9.13.

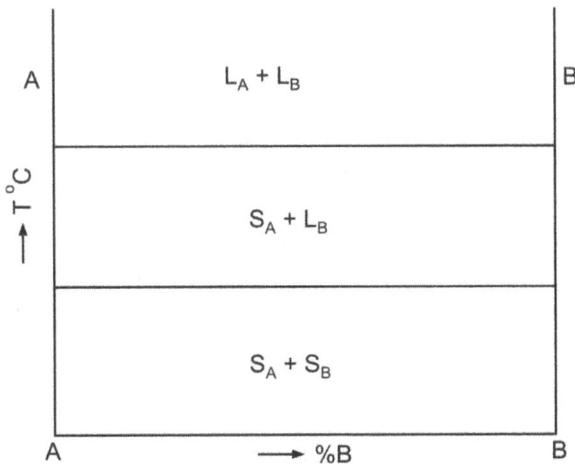

Fig. 9.13 Insoluble metals.

SOLID STATE TRANSFORMATIONS

So far we have studied different equilibrium phase changes during solidification. There are phase changes that occur after the solidifications also. We will proceed to study the changes that occur in solid state. Some important solid state transformations are:

1. Allotropy
2. Order-disorder change
3. Eutectoid reaction
4. Peritectoid Reaction.

We had studied about allotropy of metals under the crystals structures in the preceding chapters. Let us proceed to the other three transformations.

Order-disorder change: Generally this occurs in substitutional solid solutions. Ordered atom arrangement and disordered one are shown in Fig. 9.14. The solute atoms get distributed at random in the lattice of the solvent instead of occupying any well-defined specific positions, in a solid solution. This is called disorders. The reason attributed to the disordering is faster cooling which does not permits the atomic arrangement to be of any pattern. If the alloy is slowly cooled, there will be diffusion of atoms and they are formed to move into definite positions in the lattice. This is called ordering. An ordered solid solution is called superlattice.

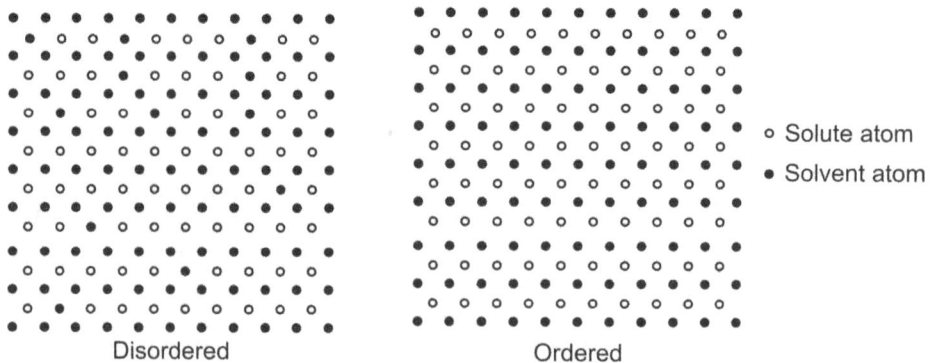

Disordered   Ordered

o Solute atom
• Solvent atom

Fig. 9.14 Atoms in solid solutions.

It is found that ordering is most common in metals which are completely soluble in the solid state. Also maximum amount of ordering occurs at a simple atomic ratio of the two elements. So, an ordered phase is sometimes represented by a chemical formula like AuCu. $AuCu_3$ in gold copper system. These phases are denoted by greek letters like terminal solid solutions i.e., α, β etc. but a stroke is put on the greek letter's top right line $α^1$ $β^1$. Gold copper equilibrium diagram is shown in Fig. 9.15.

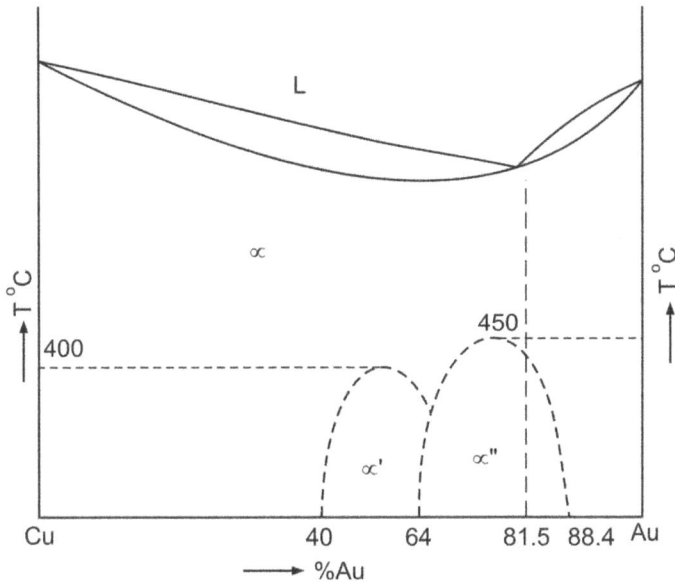

Fig. 9.15

Eutectoid Reaction: This is a common reaction in the solid state. It is very similar to the eutectic reaction but there is no liquid associated in it. One solid phase transforms into a mixture of two solid phases.

$$\text{Solid} \xrightarrow{\text{cooling}} \text{solid}_1 + \text{Solid}_2$$

An equilibrium diagram illustrating an eutectoid reaction is illustrated in Fig. 9.16. Let us study the solidification patterns of two alloys x_1% B, a pro-eutectoid and another x_2, a hyper eutecoid. In this illustration a eutectoid reaction also occurs at E besides the eutectoid at N.

When alloy X_1 is slow cooled from liquid state, solid solution y starts to form when the temperature δ crosses x_a, the liquidus. The composition of y changes along the solidus line Ta x_1b and finally it solidifies at x_b as y. Further cooling only cools y up to x_1c. At x_1c Y starts to precipitate α - solid solution, along the solidus MO while y enriches itself in B along the line X_c N. When the temperature reaches x_1d, the eutectoid temperature, r becomes unstable and undergoes the eutectoid reaction at N. Consequently the solid consists of α and eutectoid mixture. The eutectoid mixture consists of α and β. Thus the room temperature structure of the alloy is α + (α + β).

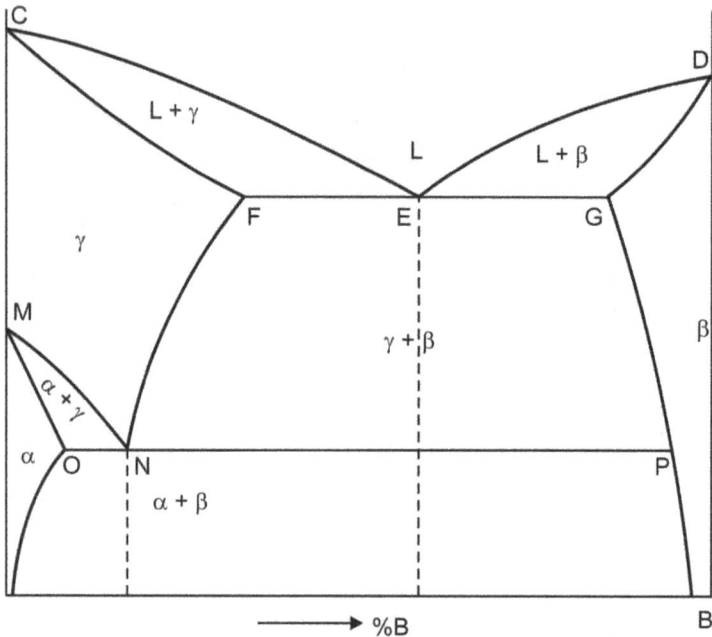

Fig. 9.16

When the alloy of x_2 is cooled from the liquid state, the solidification of γ starts at x_2a which is the liquidus temperature. The composition of γ changes along the solidus line Tax_{21} to solidify finally at x_2b. On further cooling, cools γ up to x_2c, β starts to form while the r gets depleted in B, the composition changing along the line x_2c N. When the temperature reaches x_2d, the eutectoid temperature, γ becomes unstable and undergoes eutectoid reaction giving a mixture of α and β. Thus, at room temperature, the alloy consists of β and eutectoid mixture α + β.

Peritectoid reaction: This reaction is so similar to the peritectic reaction as a eutectic reaction is related to eutectoid. It can be best understood as a reaction of the peritectic type in which the reacting liquid is replaced by another solid. It can be written as,

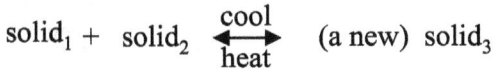

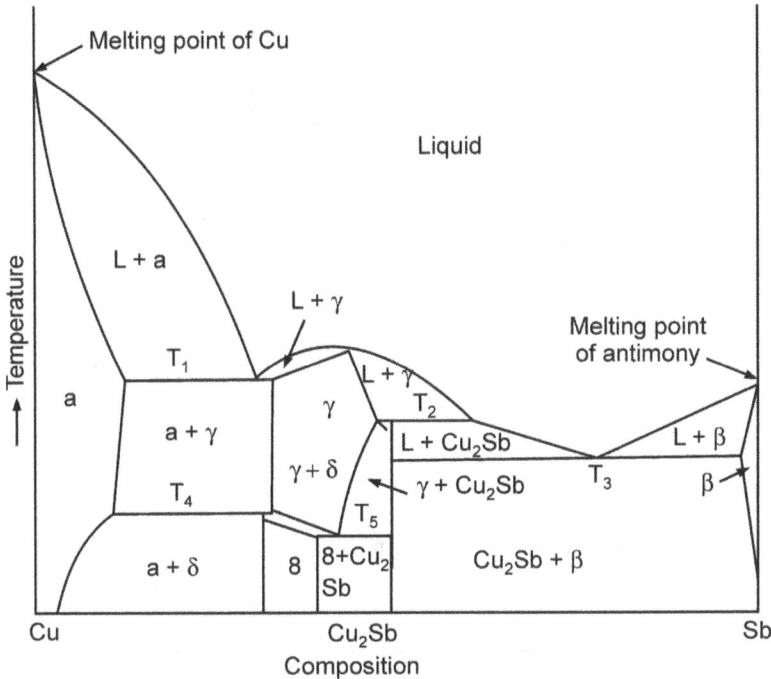

$$\text{solid}_1 + \text{solid}_2 \underset{\text{heat}}{\overset{\text{cool}}{\rightleftharpoons}} (\text{a new) solid}_3$$

Fig. 9.17 The copper-antimony system.

Generally the peritectoid reaction product is an intermediate alloy. Of course, it may also be a solid solution. A model peritectoid equilibrium diagram is shown in Fig. 9.17. The peritectoid reaction is quite common among the solid state reactions. The equilibrium reactions are listed in a tabular form, to be handy.

Equilibrium-diagram reactions

Name of reaction	General equation	Appearance on diagram
Monotectic	Cooling $L_1 \rightleftharpoons L_2 +$ solid heating	L_1 L_2 + solid
Eutectic	Cooling Liquid $\rightleftharpoons$ solid + solid heating	L solid+ solid
Eutectoid	Cooling solid $\rightleftharpoons$ solid + solid heating	solid solid+ solid
Peritectic	Cooling Liquid + solid $\rightleftharpoons$ new solid heating	Liquid + solid New solid
peritectoid	Cooling solid + solid $\rightleftharpoons$ new solid heating	solid + solid New solid

Exercise

(a) Plot an equilibrium diagram between two metals A and B with the following data;

     MP of A   -   1030 °C

     MP of B   -   720 °C

     Eutectic occurs at 410°C at 30% B.

     Solubility of B in A - 5%

     Solubility of A in B - 15% at eutectic temperature.

(b) Using the above diagram, describe the solidification pattern of an alloy A 25% B from liquid state.

Answer:

(a)   Take a common X-axis (a horizontal line) and at its both ends draw Y-axis. Mark the melting points (TA and TB) of both metals on either of the Y-axis. The X-axis denotes composition. Then, draw a light line at the eutectic temperature (FG). This eutectic line extends from 5% B to 85.0% B composition. Join the two ends of the eutectic line to the melting points. Also join the eutectic composition point (E) on the eutectic line to the melting points. Mark the phases and the axes accordingly.

(b)   To predict the solidification pattern of an alloy containing 25% B, the first step is to draw a vertical line intersecting all the phase boundaries in the diagram. This line is called Composition vertical and in this case is $X X_1 X_2$. We assume the alloy in liquid state at a temperature of X and cools from that temperature. The alloy is liquid until the temperature of X_1 is reached. At X_1, α solid solution starts to solidify and separates as α solid solution of the composition α. As the temperature is further lowered, more and more of a solid solution separates out. Its composition varies along the solidus line TAF while the residual liquid composition changes along the liquidus line TA E, when the temperature reaches 410°C (the eutectic temperature) a solid solution of the composition F will be in equilibrium with liquid of composition E. That is why the area between Ta FE is nomenclatured as (α + L).

Further fall in temperature makes the liquid of composition E to solidify as a eutectic mixture of 30% B, containing α and β phases. The composition of the solid α which had formed prior to eutectic reaction will slowly decrease in its percentage of metal B. By the time the room temperature is reached, the alloy consists of a eutectic a + b with traces of α (The solid solubility is nil at 0 °C.)

Revelation of microstructure: The microstructure of alloys is studied under a metallurgical microscope. Metallurgical microscope consists of a metal tube fitted in an arch shaped body, with a rack and pinion mechanism to move it up and down. This movement enables the observer to focus correctly on to the alloy surface under examination (Fig. 9.18). Another metal tube with an electric bulb to provide the light for illumination is attached to the microscope tube at right angles, (see Fig. 9.18). The light emanating from the bulb is focused on to the alloy specimen kept on the microscope platform, by means of a suitable prism fitted in the microscope tube. Objective (lens near the objects) is made up of suitable lens system is fitted to the tube at the bottom. The eye piece (lens near the eye) is a magnifying glass and is secured at the top of the microscope tube.

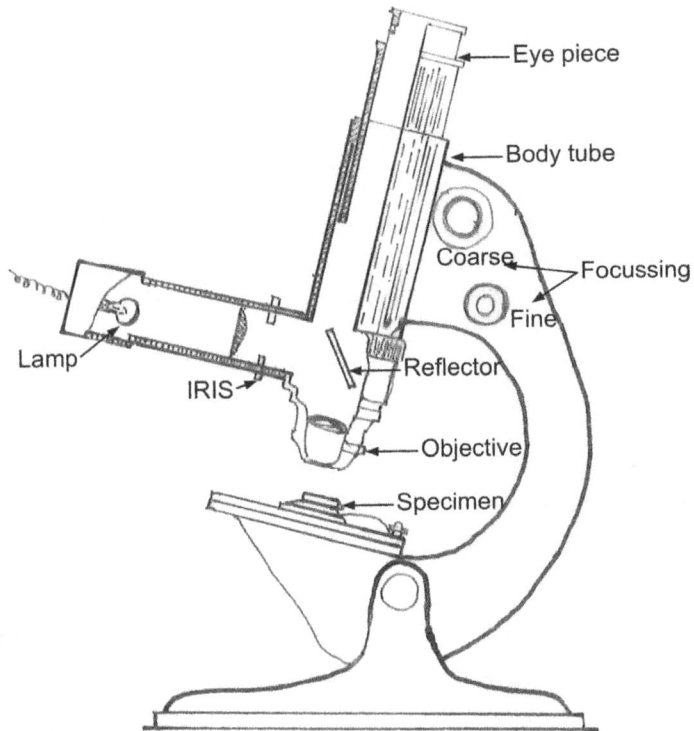

Fig. 9.18 Metallureical microscope.

Light from the light source is made to fall on the specimen vertically. It gets reflected from the surface under examination and reaches the observers eye and the surface is revealed.

The alloy to be examined is polished like a mirror to be rid of any scratches. The surface is thus perfectly horizontal and the light falling on it is just reflected back. The polished specimen looks brilliantly white. The microstructure is revealed when the specimen is subjected to a suitable acid attack. This process is called etching. Very mild etching reagents are used for a few seconds. The microconstituents are differently attacked by the etchant, leading to a micro level difference on the polished surface (see Fig. 9.19). Consequently, the light incident on the etched specimen is reflected differently. Region from which more light is reflected vertically to the eye, appears white and those regions from where some light is scattered appear dark. Appearance of a grain and grain boundary are schematically illustrated in the Fig. 9.19. It need not be emphasized that different types of phases etch differently but same phases similarly.

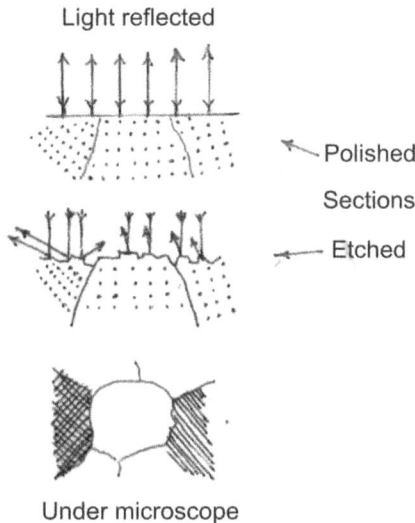

Fig. 9.19 Under microscope.

Properties of alloys: The equilibrium diagrams help us to study the various aspects of the system

   (a)   whether a solid solution is forming

   (b)   whether a eutectic occurs

   (c)   whether a eutectoid occurs

   (d)   whether any intermetallic compounds etc occur

and if so at what composition in the particular system. Having a knowledge of the different reactions, we can understand the alloy formed and remain stable at the room temperature. By having a knowledge of the probable properties of the phases to be formed, the strength and other properties can be reasonably predicted. These predictions need not be and shall not be like standard tables in a handbook, but serve a guide for proceeding with the job of suitable material selection. Further, it is difficult and in many cases impossible to predict the absolute values of the properties of a series of alloys. Some broad examples of the property evaluation in typical equilibrium systems are given hereunder.

For a system in which there is complete solubility, the hardness, density, electrical conductivity and corrosion resistance may be expected to vary.

The mechanical properties of an alloy depend upon two factors,

1. The properties of the phase or phases the alloy is composed of, and

2. The way in which different phases are existing in the structure.

In two phase alloys in which one phase is surrounded by the grains of the other, the alloy, as a whole, will have the properties of the continuous phase, e.g., Spheroidisation.

CHAPTER 10

Iron-Carbon Equilibrium

Iron exhibits allotropy. It exists with the body entered crystal structure from room temperature up to 910 °C. At 910 °C B.C.C iron transforms to F.C.C. iron which state prevails up to 1400 °C. At 1400 °C, the F.C.C. iron again changes the crystal structure to B.C.C. until its melting point of 1535 °C. The low temperature B.C.C. iron is called α-iron. The F.C.C. iron existing in the temperature range of 910-1400 °C is called r-iron and the high temperature B.C.C. iron is called δ-iron.

While heating the α-iron loose its magnetic property at 768 °C. Also it was found to absorb substantial quantity of heat at this temperature. In the earlier days, the non –magnetic α-iron was called β-iron, thinking that a phase change was occurring due to the large heat intake. Subsequently, it was established that the non magnetic iron also was made up of B C.C. crystal structure and the **nomenclature** of β-iron was dropped. That is the reason why we find the crystal structure changes called α, r and δ. Though α and δ are of B.C.C. lattice structure, there is a difference in the lattice constants of the high and low temperature irons.

Solubility of Carbon in iron: α and γ irons dissolve carbon differently. While carbon is soluble in only traces (up to 0.01%) in α iron, it is soluble up to 2.00% in γ -iron. Again, δ-iron dissolves carbon up to 0.1%. This large variation of solubilities and the formation of cementite, a compound of iron and carbon with a very high hardness have made steel a unique alloy. The heat treatment of steel has become a subject by itself (Fig. 10.1).

Iron carbon equilibrium diagram up to about 5% C is given at the top. Three phase reactions occur in this system, the peritectic reaction at 1496°C leading to the formation of solid solution of carbon in r-iron being the first. A eutectic reaction occurs between austenite and cementite at 1132 °C leading to the formation of leduberite (4.3% C.). The third equilibrium reaction is the eutectoid reaction occurring at 728 °C when austenite decomposes into pearlite. This eutectoid reaction is the 'kingpin' of the science of heat treatment of steel. Let us study the solidification of some alloys from the liquid state to the room temperature solidification of a steel of 0.15% C.

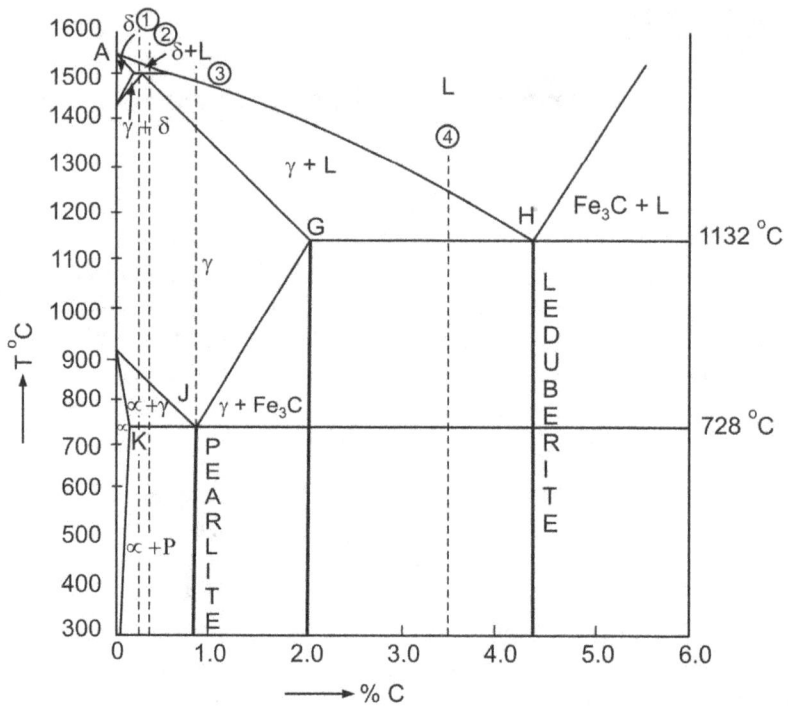

Fig. 10.1

The steel will be liquid until the temperature T_1 is reached. At T_1 δ ferrite varies along the solidus line A C while that of the residual liquid goes on changing along the liquidus line AB as the cooling progresses. When the temperature of 1496°C is reached, δ-ferrite of 0.1% C and liquid of 0.5% C react to form austenite of 0.18% C. This is peritectic reaction. So, the structure consists of δ ferrite and austenite at 1496 °C. Further cooling makes δ-ferrite to loose carbon gradually, along the line DE until the temperature T_3 is reached, when the total solid existing would be austenite. The steel cools as austenite below T_3 until the temperature T_4 is reached. A ferrite starts to separate below T_4 along the line 1K and the remaining austenite composition varies along eutectoid temperature (728°) IJ. This process continues until the temperature T_5 is reached which is the eutectoid temperature 728 °C, the steel consists of α-ferrite and pearlite. Still further cooling makes the free α-ferrite to give at carbon along the line KL, enriching the steel's pearlite content. So, at room temperature, the microstructure of 0.15%C steel will be ferrite and pearlite, see Fig. 10.2.

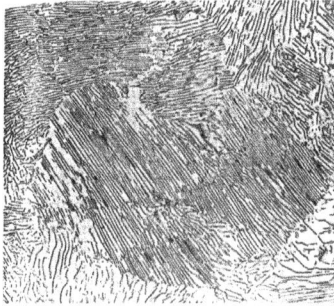

Fig. 10.2

Solidification of a steel of 0.3% carbon: The steel is liquid above T_1 solidification starts at T_1 with δ-ferrite. The composition of δ-ferrite changes along the line AL, while that of the remaining liquid along the line AB. This continues until T_2, the peritectic temperature. Peritectic reaction occurs with some δ-iron of 0.1% C reacting with the liquid of 0.45% C to form austenite of 0.18% C at T_2. Below T_2, the solid existing is austenite and some δ-ferrite denoted by the field CED. Further cooling results the microstructure the steel undergoing changes, similar to a pattern described in the previous example of solidification of a steel containing 0.8% carbon.

The steel is liquid above T_1 solidification starts at T_1 with the separation of austenite whose composition varies along the line DF and that of the residual liquid along the line BG. By the time the temperature reaches T_2 the whole liquid solidifies as austenite. Between T_2 and T_3, austenite only cools. At T_3, austenite undergoes the eutectoid reaction and forms pearlite. Thus, a steel containing 0.8% carbon consists of full pearlite at room temperature. Because of the eutectoid reaction the steel undergoes, 0.8% carbon steels are called eutectoid steels.

Solidification of an alloy containing 3.0% carbon: This alloy contains carbon in excess of 2.0% and so falls into the irons category. Solidification starts T_1 with the separation of austenite. The compositions of austenite separating and the residual liquid vary along the solidus DF and liquidus DH respectively. At T_2, austenite containing 2% carbon and liquid of 4.3% carbon will be in equilibrium. Eutectic reaction occurs at this temperature and the liquid solidifies as a eutectic called leduberite. But, as cooling is continued below T_2, austenite starts to give out excess dissolved carbon and a compound cementite forms. Again, at T_3 the balance austenite (of 0.8% C) undergoes eutectoid reaction and forms pearlite. Thus, at room temperature, the cast iron at 3.0% carbon consists of pearlite, leduberite and cementite.

STEELS

We have studied about iron-iron carbide equilibrium. We have also studied the different carbon steels. Just to recapitulate, we should understand that all the iron-carbon alloys containing up to 2.00% C are steels. In actual practice, steels contain carbon up to a maximum of 1.3 per cent. The different steels and their microstructures were also discussed when we described the iron-iron carbide equilibrium.

Classification of Steels: Carbon steels are those steels containing not more than 0.5% manganese and 0.5% silicon and all other steels being regarded as alloy steels.

Alloying elements added to steel vary widely in amounts either singly or in complex mixtures. The purpose of alloy addition is to obtain one or some properties in an improved range Alloy addition is not a ready made solution to get a steel of all perceivable improvements in mechanical properties at one stroke. In certain cases, the use of an alloying element may intensify rather than diminish the difficulties present. So, the engineer should properly understand the underlying principles involved in the alteration of the characteristics of the steel by the addition of the alloying elements. He should also be familiar with the general effects of various alloying elements, that will enable him to spot the situation correctly when an alloy steel will suit him for the design.

Addition of different alloying elements to the steel brings about changes in the equilibrium. However, no new phases will appear, though the composition and proportion of the phases in the structures may be altered considerably. Allloying elements may also modify the transformation temperatures and thus modifying the properties of the plain carbon steel. Thus it should be borne in mind that the entire effect of the alloying elements is to modify or enhance the existing properties of the plain carbon steels.

Generally, the effects of alloying elements in steel can be one or some of the following ways,

1. May form solid solution or intermetallic compounds
2. May alter the temperatures at which phase transformations occur
3. May alter the solubility of C in r and α iron
4. May alter the rate of reaction of transformation of austenite to its decomposition products. Similarly the rate of solution of cementite into austenite while heating.

5. Alloying elements may decrease softening of the steel while tempering

6. Alloying elements may strengthen the steels which cannot be quenched.

Practically, all alloying elements added to steel viz, Ni, Si, Al, Mn, Cr, W, Zr, Mo, Ti, P, S and Cu are soluble in ferrite to varying extents. Some of these elements form carbides when sufficient carbon is present. Sometimes, non-metallic products may be formed which localize in the structure as inclusions. Intermetallic compounds are formed in some cases. The first seven elements do not form carbides. Si, AL and Zn however, tend to oxidize and from oxides. Silicon and aluminum are known deoxidizers in steel making. The manner in which steel is de-oxidized influences the austenitic grain size e.g., when aluminum is used as deoxidizer, the steel will have a fine grains size. Aluminum trioxide formed is finely distributed in the structure to promote nucleation and hence the fine grain size results. The elements, Mn Cr, W, Mo, V and Ti have stronger tendencies to form carbides in that order, Ti having the greatest tendency. However in the absence of carbon, these elements dissolve in ferrite to a certain degree, the lower ones in the series has more tendency to form carbide than the higher ones

Phosphorus in quantities generally found in steels, dissolves in ferrite with considerable ease. However, when present in larger quantities, it forms iron phosphide which promotes brittleness in steel. Sulphur forms iron sulphide which locates itself at the ferrite pearlite grain boundaries. This reduces the ductility of steel at forging temperatures. This is called hot shortness. Manganese is added to minimize this draw back. In the presence of manganese, an iron manganese sulphide is formed which get uniformly distributed in the structure, thus is less harmful than iron sulphide. Copper does not form carbides. It dissolves in ferrite up to 0.8%. This limited solubility of copper is made use of in some cast copper bearing steels to better the strength properties, the rough precipitation hardening.

The addiction of alloying elements alters the temperature at which the γ-iron changes to α iron. Similarly the eutectoid temperature may alter; either increase or lower the temperature of transformation critical temperature may be raised while heating and lowered while cooling. Another significant factor is a change of eutectoid composition due to the addition of some alloying elements. The curves given illustrate the effect of alloying elements on the eutectoid temperature and on the eutectoid carbon percentage. Thus, with proper alloying mix, steels with austenitic structure are made at room temperature. These are commonly known by the name stainless steels because they are corrosion resistants (No stain or mark forms on the steel i.e., corrosion) (Fig. 10.3).

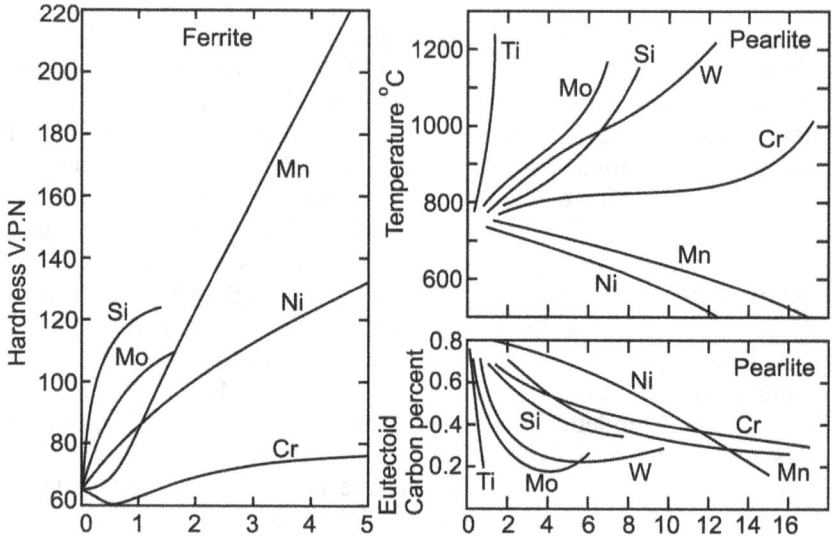

Fig. 10.3

Classification of plain carbon steels: Alloys of iron with carbon up to 2 percent defined as steels. Yet, the practical range of steels is only up to say 1.3% carbon. These steels are divided into three classes.

1. Low carbon or mild steels	up to 0.25% C
2. Medium Carbon steels	0.25 to 0.5% C
and 3. High Carbon Steels	0.5 to 1.3% C

1. **Low Carbon Steels**: As said above, these are popularly known as mild steels. Bulk of the steel produced is taken up for construction works – fabrication and manufacture of house hold items like nails, clamps etc.

2. **Medium carbon steels**: Many forging steels fall into this category. Rolled products like plates used in the construction of engineering equipment like boilers, machinery bodies, heavy duty steels nuts

and bolts etc make up use of this steel Rolled into engineering sections of numerous shapes in erection work.

3. **High Carbon Steels**: In the high carbon steel category come the most important rolled product – rail road rails, wheel centers for railway wagons are forged with this varity of steels. Steels, of the higher carbon ranges are used for engineering tools for cutting and slicing.

In fact, there is no demarcation that steels of this percentage carbons are used for this purpose. Trace alloy addictions are found to do wonders with respect to the performance of the steel in service. Fig. 10.4 gives the hardness of steel with changing carbon content. The different properties vary with increasing carbon is graphically illustrated. Some typical compositions and their properties are given in the table 10.4 for easy understanding.

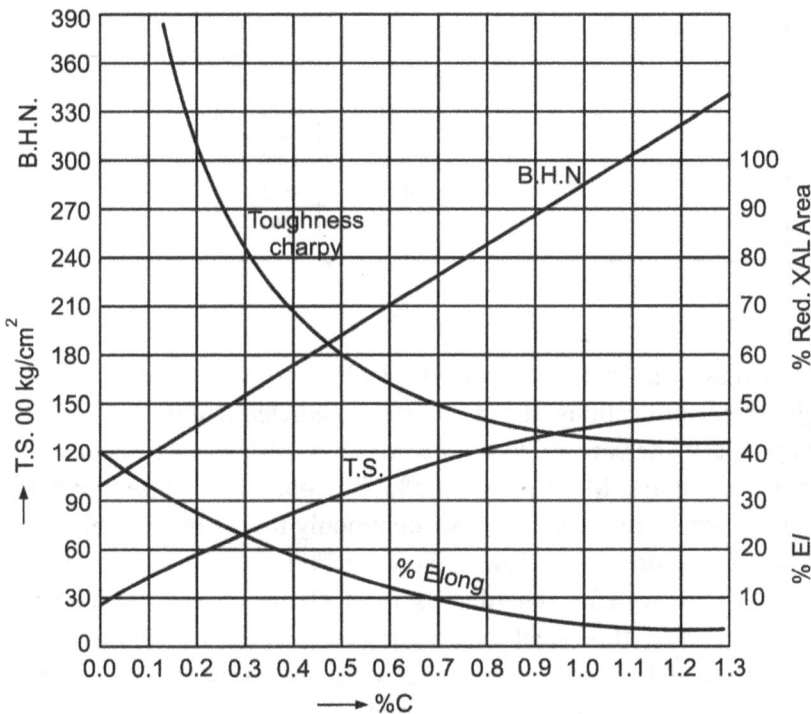

Fig. 10.4

PLAIN CARBON STEELS

Composition					Conditions	Y.P	T.S.	EL	H
C	Si	Mn	S	P		Kg/Cm^2	Kg/Cm^2	%	B.H.N.
0.14	0.2	0.56	0.03	0.03	N.910°c	3100	6050	44	117
0.15	0.09	1.1	0.28	0.08	Hot rolled	3800	6050	40	110
					Cold Drawn	4480	6950	24	155
0.2	0.25	1.0	0.03	0.03	Casting	6030	6950	18	149
0.24	0.1	0.55	0.04	0.04	A880 6Hrs	6030	6720	30	147
					N.870	3600	7715	34	134
0.35	0.15	0.7	-	-	0.Q 860 T650	4480	8060	36	134
0.58	0.37	0.79	0.04	0.03	Forging 0Q850	4480	8300	23	164
					T600	6050	9630	27	190
0.65	0.25	0.76	-	-	N 850	6720	11200	18	220
1.15	0.02	0.32	0.03	0.03	0Q830	7840	12770	20	250
					T600 As rolled	6720	11200	22	-
1.56	0.03	0.18	-	-	0 Q820 T600	8290	13900	17	275
					As rolled	8512	12550	7	-
						8064	10750	2	-

Silicon Steels

Silicon dissolves in ferrite. Silicon steels are quite commonly used in engineering applications. It can be safely predicted that more than 70% of the springs made, ore of low silicon variety. This steel contains 0.5% carbon and 0.8% Mn besides 1.5% Silicon. Cold coiled springs of numerous sizes are in use. Another commonly used silicon steel is named as transformer steel. Very low carbon (0.07%) and manganese present in this steel may well be regarded as most elements. With silicon in the range 4.5-5.5%, this steel has a unique property of high magnetic permeability and electrical resistance. Main use, as mentioned before is in the manufacture of dynamo poles, and stampings for the transformers. Another alloy known as silichrome containing 0.4% C, 3.5% Si and 8% Cr is used for the manufacture of valves in auto engines. In general silicon steels possess good oxidation resistance.

Nickel Steels

3.5% Nickel steels of low carbon are extremely used for carburizing of drive gears, connecting rod, bolts, studs, and kingpins. 5% nickel steels provide increase toughness and so are used for bus and track gears coins, and crankshafts. Nickel has great ability in improving the toughness of the steel.

Chormium Steels

Chromium being cheaper than nickel and has a high solubility in y iron. In low carbon steels, chromium goes into solution thereby increasing the strength and toughnes Chromium in quantities excess of 5% imparts high temperature properties and corrosion resistance to the steel. But, to impart greater toughness, a small quantity of nickel is also added. In fact, when chromium and nickel are used in the steel, they will be in the ratio of 2:5:1 nickel & chromium. Nickel's toughness quality coupled with improving hardness by chromium come handy. Low carbon nickel chromium steels (1.5% Ni and 0.6% Cr) are used for worm gears, piston pins etc. The nickel contents is increased to 3.5% and chromium to 1.5% for the manufacture of heavy duty item like automotive connecting rods and cams.

Manganese Steels

Manganese is one of the cheaper but effective alloying elements. Only when the percentage of manganese exceeds 0.8% the steel is classed as alloy steel. Like nickel it lowers the critical range and eutectoid carbon percentage, fine grained manganese steel attains vey high toughness and strength. So they are also used for gears, spline shafts, axles and rifle barrels. Manganese steels, with a small amount of vanadium added are used for large forgings which are air cooled.

When the manganese content exceeds 10%, the steel will be austenitic on slow cooling. The well-known Hadfield manganese steel contains 11-14% Mn and 1.1-1.4% C. After a properly controlled heat treatment, the steel develops high strength, high hardness and ductility besides excellent resistance to wear. Hadfield manganese steel is 'the material' for severe service conditions that warrant abrasion and wear.

When this alloy is austentised at 1150 °C and quenched, tensile strength up to 12000 kg/cm^2 is developed with an elongation of 45%. It is tempered at around 300 °C to relieve the quenching stresses. When it is pressed into service involving sudden and repeated loads, its hardness dramatically increases to 500 BHN. This is due to the conversion of source of the austenite to martensite. It is used for earth moving equipment showels, bull-dozers and rail-line points due to this property.

Stainless Steel: Stainless steels are used both for corrosion resistance and high temperature uses. They are generally divided into three types

> 1. Maternsitic
>
> 2. Ferritic
>
> and   3. Austenitic

Of the above, the third variety, austenitic stainless steels are very common. These steels are non magnetic in the annealed condition. The common composition is 18/8 i.e,Cr/Ni with very very low carbon. They can be cold worked and also hot worked and have excellent high temperature strength and resistance to scaling. This steel is commonly used in the manufacture of cutlery items.

Tool Steels

Selection of a proper steel for a particular tool for a given application is a delicate job. The best way to approach is to have a thorough understanding of the metallurgical characteristics of different steels vis a vis the requirements of the tool. No tool and steel is earmarked for a particular job nor it can be earmarked for a particular set of operations as a working solution. Still, it is gratifying to note that many tool steels will perform any given job reasonably well. The steel to be used for a particular job is judged by the performance, ease of its fabrication (into the tool) and cost. Here cost means the cost per unit part made by the tool. These three criteria determine proper steel selection.

Except those steels that go into the production of a machine part, other tool steels are mainly divided into the type of operations – the tool is going to perform, cutting, shearing, forming, drawing, extrusion, forging and rolling. A cutting tool may have a single cutting edge like the one used in a lathe or planer or may have two or more cutting edges which perform continuously just like a drill, tap or each edge taking short cuts and functioning only a part of the time like milling cutters or hobbing. These steels should not only possess high hardness but good wear and heat resistance.

Shearing tools should possess high wear resistance with a reasonable degree of toughness. These two properties should be properly balanced, keeping in view the operation performed, like the thickness of the stock and the temperature of the shearing operation.

Forming and forging dies should have high toughness and strength besides 'red hardness'. Drawing and extrusion dies must possess high wear resistance and high temperature properties (red hardness), besides strength, to withstand the pressures.

Thread rolling dies must be hard and withstand the pressures generated in forming the threads. Battering tools like hammer heads and chisels should possess high strength and toughness.

From the above, high strength, hardness, wear resistance, toughness and red hardness are the important properties to be considered in all the applications while selecting steel for the dies or tools.

Red-hardness: When a machine operation is going on, the cutting tool will be removing the metal from the surface of the revolving stock. There will be a lot of friction between the tool tip and the stock surface. Large quantity of heat is generated. Consequently the tool becomes so much heated up that it becomes red. Previously, when carbon steel tools were used, the tool used to loose its hardness and become soft in that heat condition. The tool gets so hot that it turns red. A better tool made of the steel which will not soften even at higher temperatures of say 700 or 800°C, though it turns red while in use. This property of retaining its hardness even at red - heat, is termed red-hardness. In fact it is a requisite property of any turning or machining tool.

Non deforming properties: Steels expand and contract during heating and cooling. These volume changes are detrimental to the die. Dies of metal working are of complex shapes and once even a small change occurs in them, the final product will be of a different dimension. So, it is imperative that the die maintains its dimensions both during its manufacture i.e., heat treatment etc. and during service. Steels which are good with respect to non-deforming properties can be machined close to the final size before the heat treatment.

In general, air hardening steels exhibit least distortion. Steels of oil quenching and water quenching types are prone to distort.

Toughness: The term toughness as applied to tool steels should be thought of as the ability to resist breaking rather than the ability to absorb energy during deformation. All the tools must be rigid as even the slightest deformation makes it unfit for use.

Wear Resistance: Wear resistance is defined as the resistance to abrasion or loss of dimentional tolerances. Several tool steels are excellent with respect to this property.

Red hardness: It is also called hot hardness and related to the softening of the steel at high temperatures. This is reflected to some extent, by the resistance of the material to tempering which is an important selection factor for high speed and hot working tools.

Machineability: When dies are made out of the steel, it should cut easily. Many dies are of intricate shapes which warrant various workshop operations to manufacture. The steel should possess the quality of easy cutting, without leaving tool lines or tool marks on it.

Resistance to decarburization: Almost all the tools and dies are heat treated before putting into use. Decarburisation is the loss of surface carbon on heating. It occurs usually at temperatures above 600°C. When the steel looses some of its surface carbon it becomes soft, instead of becoming hard after quenching. So, decarburisating tendency should not be there in the tool steels

Shock resistant tool steels: These steels are generally tough with ability to withstand repeated shocks (forging dies). They have very low carbon (0.45-0.65%) and alloying element Si, Cr, W and sometimes M_o.

High Speed Steel: These are among the most heavily alloyed tool steels and usually contain tungsten or molybdenum along with chromium and vanadium, sometimes cobalt. The carbon content ranges from 0.7 to 1.0 percent. These are subdivided in to tungsten base and molybdenum based steels. The most widely used tool steel contains W, Cr and V, in percentages of 18, 4 and 1 respectively. They are named 18/4/1 steels for brevity.

Heat treatment: High speed steel ingot structure is similar to cast iron except, instead of cementite, it has $(Fe\ W\ Cr\ V)_6C$. This extremely brittle network is broken up into small globules by annealing and forging, of course avoiding laminations of carbides.

It is annealed to soften at around 850°C for above 4 hrs. During annealing, it is protected against oxidation. Tools are forged to the required shape. Then, they are heated to 680°C for ½ hr and air cooled. The structure consists of carbide globules in a fine pearlite matrix.

Hardening: Y(austenite) forms on heating at 800°C, but contains only 0.2% carbon in solution. Quenching produces Fig. 10.13 martensite (Fig. 10.13). More of carbides dissolve on holding as indicated by the line

EB. Quenching produces structures of increasing red-hardness due to the presence of a large percentage of alloying elements in solution. This makes the steel sluggish to tempering. Even at a temperature of 1300°C when melting occurs, only about 0.4% carbon dissolves in austenite, the balance remaining as complex carbides. So, to attain the maximum cutting efficiency, sufficient carbon as well as alloying elements must be dissolved in the austenite. This necessitates a temperature little short of melting, say 1150-1250°C. At this high temperature grain growth and oxidation are prone to occur rapidly. So, the tools are first carefully preheated to 850°C soaked and then rapidly heated to the hardening temperature and quenched in oil. Instead of oil quenching, they can be cooled in an air blast. The following modification may be made to minimize the severe thermal stresses set in whole quenching, the temperature gradient from the surface to the core is minimized by

(a)   Cool in a salt bath at 600°C, homogenize and then quench in oil

(b)   Oil quench at 425°C and then air cool.

Tempering: High speed steel contains a lot of retained austenite. So, in this quenched condition itself they are less hard than fully martensitic steel. The retained austenites are dissociated by tempering or sub-zero cooling to – 80°C. If tempered, it is found that tempering more than once is more effective than a single temper (of the same duration).

Tempering 350-400 °C: The toughness increases, though there will be a slight decrease in Hardness.

Tempering 400-600 °C: Hardness increases, so this tempering treatment is sometimes referred to as secondary hardening. Of course the structure is made up of fine spheroidal carbides in an austenite martensite matrix.

Mechanical Properties: Transverse Strength - 50,000 Kg/cm^2

Compression Strength - 60,000

Modules of Elasticity 320,000 Kg/cm^2

Hardness Ra – 85

Knoop – 740

Note: The tensile strength and hardness of steels is related by an empirical formula known as Beeching's formula which is as follows: T.S.(ton/cm^2) = K × BHN

where K is a constant whose value is 0.22 for plain carbon steels and 0.217 for alloy steels.

Effect of alloying elements on Fe-C System

We have discussed the effect of different alloying elements on the properties of steels, in the previous chapter. There, the discussions were reasonably elaborate and of course, concentrated on the effect of various alloying elements on the general sense. Let us proceed to study the manifestation with respect to the changes that are brought about in the equilibrium diagram from a different angle.

We know that some alloying elements raise the eutectoid temperature while some lower it. Again some alloying elements alter the carbon percent where the eutectoid reaction occurs. These changes have their own effects on the properties of steel. To be more peruse nickel and manganese by lowering the eutectoid be "quenched" from a lower temperature than an ordinary (plain carbon) eutectoid steel. So lesser strain is experienced illustrated in Fig. 10.5.

The effect of alloying elements like chromium, vanadium, tungsten, molybdenum, silicon and titanium will be to shift the critical temperature to higher values. Such a shift will bring the delta ferrite to different extents but the result is same.

Consequently, the austenitisation of steels containing such elements is possible only over a narrow range of carbon percentage. This is a practical difficulty for any heat treatment process is started with austenitisation of the steel only.

Critical cooling rate: Some alloying elements effect the critical cooling rate of steel. This effect is to shift the critical cooling rate come to the left on to the right. When the critical cooling curve is shifted to the left, the steel will harden only by very fast cooling and vice versa. We will discus more about this aspect at the relevant place is hardening. However the effect of some elements on the critical cooling rate is illustrated in (Fig. 10.6).

°C   Eutectoid temperature A$_1$

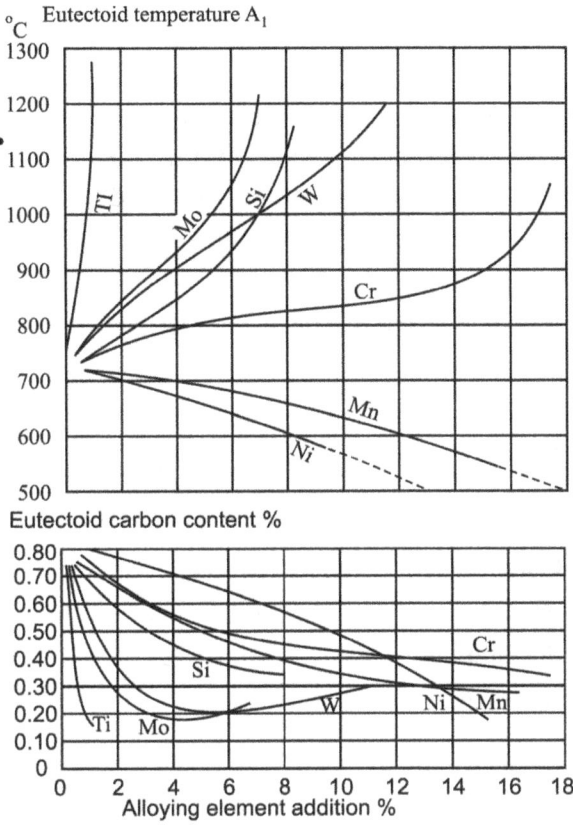

Eutectoid carbon content %

Alloying element addition %

Fig. 10.5

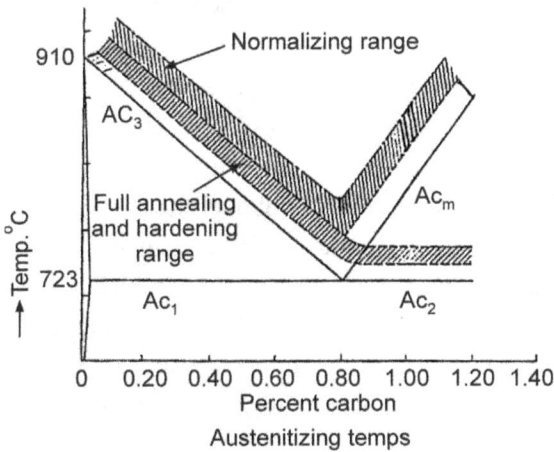

Austenitizing temps

Fig. 10.6

ANNEALING

Annealing is defined as a process of heat treatment in which the material is heated to a pre determined temperature, soaked at that temperature depending upon its size for a certain time and then cooled slowly. Annealing is done on steels (and many other materials) to accomplish any or some of the following,

1. Stress relief and/or grain refinement after cold working.
2. Stress relief in castings arising out of different cooling rates.
3. Induce softness and machineability.
4. Improve grain structure viz., grain refinement.
5. Homogenization (of composition)
6. Remove dissolved or entrapped gases.
7. As a special treatment like malleablisation in cast irons and spheroidisation in tool steels, and
8. Control mechanical properties.

Temperature of heating

The temperature to which the steel is heated and soaked depends up on the purpose for which annealing is done. The annealing temperature also depends upon the carbon percentage of the steel. It is pictorially shown in (Fig. 10.6). The temperature of annealing will be above the Acm line to the effect (Fig. 10.6) homogenization of the structure. It is so because; all the cementite and other carbides should be taken in to solution by the austenite. Similarly, for grain refinement and spheroidisation the temperature of heating should be above the A, temperature. But the temperature of heating of the steel will be around 300°-350° C in stress relief annealing. This process is also called sub–critical annealing as it is performed at a lower temperature than the critical temperature .When it is performed intermittently during a metal working process like wire - drawing, it is known as process anneal because it forms a part of the process of manufacture.

Malleablisation, spheroidisation and removal of entrapped gases, require annealing at temperatures higher than the transformation temperature (of austenite) i.e., above A_3 and Acm, depending up on the carbon content.

Soaking

Soaking is the second stage in the annealing procedure. The object or objects being annealed are held for some time at the temperature to which it/ they are heated. The necessity of soaking is to make the temperature of the object uniform throughout its section. The surface and some shallow layers under the surface attain the temperature of the furnace almost instantly. But, the stock is bulky the heat of the furnace should be conducted into the core portion of the stock by conduction only and many, steels and alloys are not so good conductors. So, it takes some times for the temperature to get uniform throughout the stock. Hence this soaking stage has arisen. The time of soaking is generally determined by the thumb rule calculation – one hour per inch of section. At the end of soaking the structure of the steel over its entire section would be austenite. That is why, many call the heating and soaking periods together as 'austenitisation'.

Cooling

The annealing process is complete when the soaked steel is cooled slowly to the room temperature. Slowly is a vague term. By definition, slowly is associated with "equilibrium' cooling. Equilibrium cooling in letter is a meaningless expression – because when cooling occurs in the temperature changes, where is equilibrium? But, "de facto" equilibrium cooling refers to very slow cooling. Some scientists suggest that the stock of left in the hot furnace itself and the heat put off, so that it cools along with the furnace. This is impracticable even on a laboratory scale. So the objective is achieved by cooling the stock in sand or lime – the idea behind is of course, very slow cooling.

Some microstructures are given in Figures 12.6 to 12, which are self explanatory. The way the microstructures are changed to grains of uniform size and equi axial texture by the process of annealing is seen from these structures. Of course, how the carbon is made to modify its existence from combined state to spheroids can be seen from the spheroidcal graphite cast iron structure. Structure of a 1% carbon steel with it cementite network broken up and spheroid zone is shown in (Fig. 10.7). This improves the machining qualities of the steel.

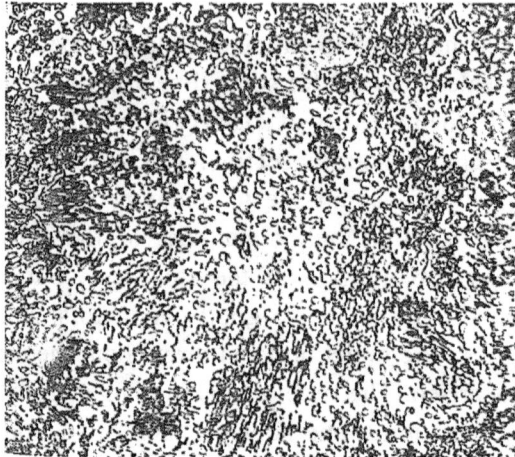

Fig. 10.7

CHAPTER 11

Normalising

Normalising is a heat treatment process akin to annealing. In normalizing, the steel is heated to the austenitising temperature, let it be of any carbon content. If the carbon content is more than 0.8% it is austenitised above Acm temperature, unlike in annealing and socked and cooled in still air. Thus normalizing differs from annealing only with respect to the austenitising temperature but also with regard to cooling.

Still air cooling makes the cooling to occur at a relatively faster rate. The steel possesses a comparatively finer grain structure.

Normalising treatment is many times preferred to annealing for its relative quickness and because a normalized steel possesses a finer grain size. Many case carburring, nitriding and sometimes hardening procedures prescribe the normalizing of the steel prior to the actual processes. Even otherwise, a normalised steel will have a higher strength and toughness, in comparison to an annealed steel of the same composition.

DECOMPOSITON OF AUSTENITE

We have seen that for any heat treatment the steel has to be austenitised. Whatever the constituents of the steel using be at the room temperature whether they be ferrete and pearlite, or pearlite or peralite and cementite, all will be disappearing and form the phase austenite when it is heated to a temperature, above the critical temperature. This is because, austenite is the solid solution of carbon in Firon and Firon can dissolve carbon up to 2.00%. The eutectoid steel has the lowest transformation temperature of 723°C. This, eutectoid steels are austenitised at temperatures around 750°C. When the steel is cooled from the austenite region, it again forms peralite due to the eutectoid reaction.

When the austenite is cooled, a small nucleus of cementite forms at the grain boundary once the temperature reaches this eutectoid temperature. When this occurs, the neighbourhood of the nucleus becomes depleted in carbon (cementite carbon 6.67% C whereas austenite at that composition had only 0.18% C). This region which is depleted in carbon automatically becomes ferrite. Cementite nucleus grows longitudinally as it is easy to garner carbon from either side than the austenite which is slightly away. Another nucleus and its longitudinal growth, with ferrite on either side. The nucleation and growth of cementite in conjunction with ferrite as alternate plates is illustrated in Fig. 11.1.

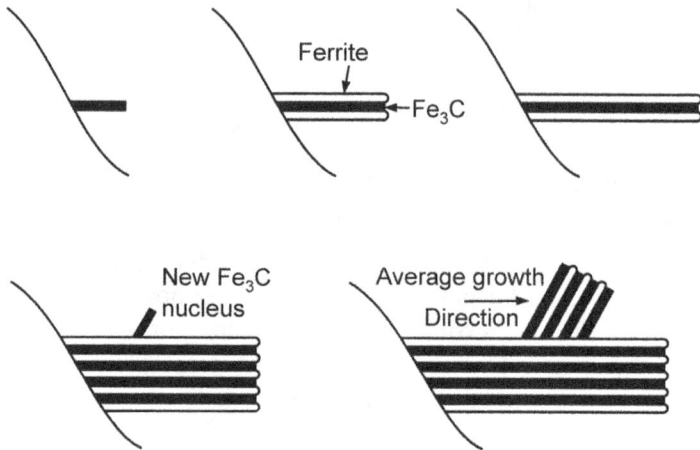

Fig. 11.1 Nucleation and growth of pearlite lamellae.

Migration of carbon from austenite into cemtite for its growth needs sometime. During equilibrium i.e., slow cooling, carbon can migrate relatively larger distances than during faster cooling. That is why, we get a fine pearlite structure on faster cooling and course pearlite by slow cooling. It should also be noted that if the austenite is fine grained, there will be more grain boundary area, more possibility for cementite nuclear to form and thus more lamellae – a fine grained pearlite results.

Since diffusion is a rate process (time dependent process), it is possible to suppress the eutectoid temperature. However, at these subcritical temperatures, austenite is metastable (stable only for a short period) and tends to decompose into the ferrite-cementitc aggregate – pearlite or other structures, if sufficient time is allowed. The study of this subcritical transformation of austenite under isothermal conditions has resulted in the so called T.T.T (Time-Temperature-Transformation) diagrams or C-Curves or S-Curves.

In order to study the isothermal transformation of austenite at subcritical temperatures, say of eutectoid steel, a number of thin pieces of that steel are austenitised at 750°C. Then these specimens are quenched into a lead bath or salt bath which is kept at a temperature at which the transformation is to be studied. The specimens are allowed to transform for varying periods and at regular intervals, the specimens are withdrawn one by one, and quenched in water. Those specimens in which the transformation has not started as yet, the structure after quenching is completely martensitic. Those specimens, which are kept for a period long enough for the austenite-pearlite transformation to be completed, the structure shows 100% pearlite. The specimen in which 1% pearlite and 99% martensite is seen, must have been kept for that much time which is the minimum time required for the starting of austenite to peralite transformation. Similarly a specimen in which martensite is less than 1% must be that which remained at the isothermal transformation temperature for a duration which was just sufficient to complete the austenite-pearlite transformation. Thus the time required for the initiation and completion of austenite-pearlite transformation at different temperatures can be found out by repeating this experiment at different temperatures. When these results are plotted graphically, on a time-temperature diagram, and the times for the start and that for completion, are respectively joined by smooth curves, we get two curves which have a shape resembling the letter C or S. Hence the name. The curves are shown in Fig. 11.2.

Time Temperature Transformation Curves

The time required for austenite-pearlite transformation to start, is known as the incubation period.

On quenching, the austenite to various temperatures, it has been found that the martensite does not start to form unless the austenite is cooled below a certain minimum temperature which is characteristic of a given steel and depends mainly on its chemical composition. Similarly the martensite transformation is almost completed on cooling the steel to a further lower temperature. These temperatures are termed as Ms and Mf temperatures respectively and are characteristic for a given steel. These temperaturesare also shown in Fig. 11.2.

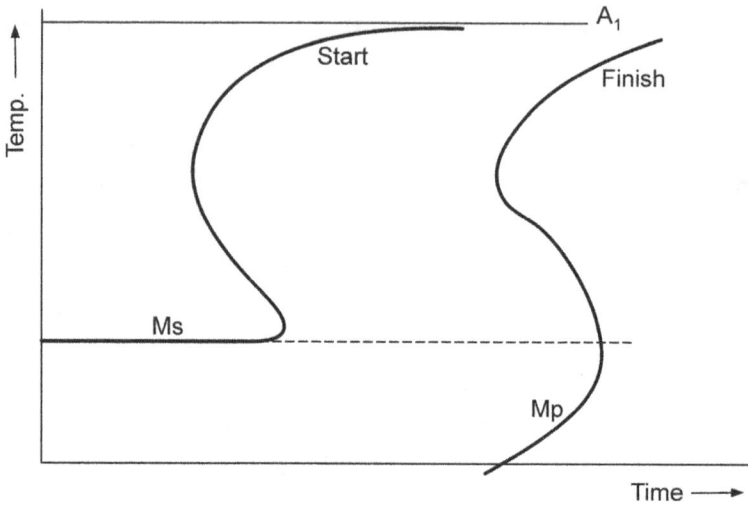

Fig. 11.2 Time temperature transformation curve.

When austenite is isothermally transformed at subcritical temperatures above the Ms temperature, it decomposes into a variety of structures, all of which are ferrite-cementitic aggregates in nature, but differ in fineness of dispersion of the two phases. The names of the phases obtained by the transformation of austenite at various temperatures are labeled in the Fig. 11.3.

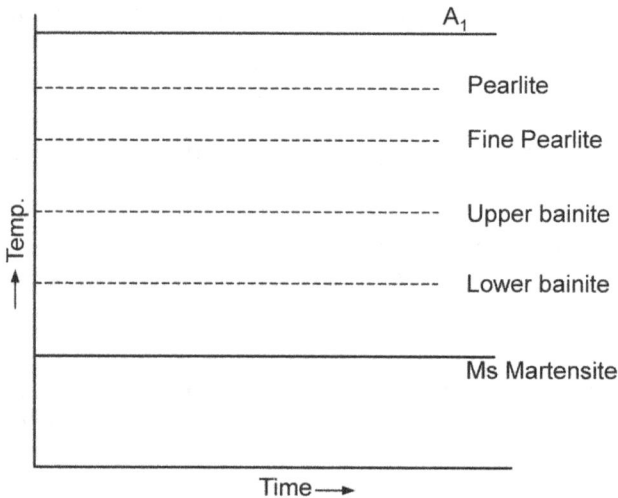

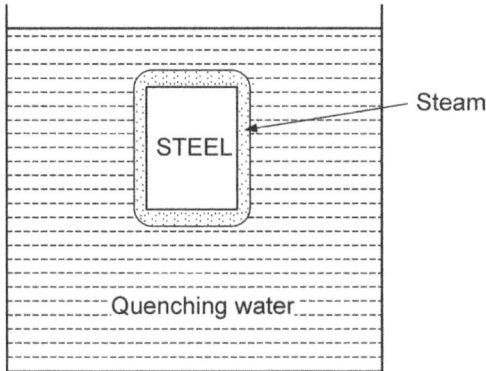

Fig. 11.3

At higher temperatures, since the diffusion of carbon is faster, and the nucleation rate slower, the lamellae of ferrite and cementite can grow to large dimensions and as such the interlamellar spacing of pearite increases resulting in coarse pearlite. At temperatures near or just above the nose of the C curve, the incubation period is smaller, indicating a high rate of nucleation, i.e., a large number of nuclei. Since the temperature is lower, the mobility of the C atoms is decreased and as such the ferrite and cementite lamellae can not grow to large dimensions. Thus they are very fine and the interlamellar spacing is small. This structure is called fine pearlite. Since the disperaion is finer in this case, the mechanical properties are superior to that of coarse pearlite.

Isothermal transformation of austenite below the nose of the C-curve gives rise to structures which are formed entirely by new mechanisms, although they are essentially ferrite-cementitic aggregates. These structures are termed as upper bainite and lower bainite. The dispersion is very fine in both the cases. As such these structures are very strong, hard and tough. Lower bainite has the optimum combination of high hardness and toughness. It is also known as acicular bainite because of its needle-like structure resembling to that of martensite, while upper bainite, because of its feather like structure, is also called feathery bainite. Ferrite and cementite in these bainites can be seen only at very high magnifications.

Non-Equalilibrium Cooling

It is seen from the TTT diagram, that the transformation of austenite to pearlite is suppressed and martensite can be formed instead, by cooling the steel at rates faster than the critical cooling rate - represented by the cooling curve which is a tangent to the S-curve.

Higher cooling rates are achieved by the process of quenching i.e., sudden cooling. This is attained by employing water, oil, salt water (brine), compressed air and even solid carbon dioxide, depending upon the necessity.

Mechanism of quenching: The Fig. 11.3 shows a steel austenitised and dropped into water. The various stages of its 'chilling' can be analysed and classified into three names - vapour blanket stage, vapour transport stage and liquid cooling stage.

When the hot steel comes into contact with water, water immediately forms steam. The steam sits on the surface of the steel and envelops the steel as a blanket. That is why this stage is named as vapour blanket stage. More and more vapour forms and accumulates on the existing 'blanket' only until its thickness becomes so much that it cannot continue to stick. Then, the vapour blanket breaks up and the steam (vapour) starts to bubble up and escape out. This is the beginning of the second stage-called vapour transport stage, not only the vapour present as the blanket escapes out but also the freshly formed vapour, (as the steel surface starts to came into direct contact with the liquid) also bubbles out. This continues until the temperature difference between the steel and the liquid decreased to such a low that no more steam (or vapour) forms instantly. The third stage, the liquid cooling stage sets in when the cooling is supposed to follow the Newton's Law.

The cooling during the first stage i.e., the vapour blanket stage is very slow as the steel is physically separated from the cooling medium by the vapour blanket. During the second stage, the cooling rate is highest and during the third stage is a quite considerably and steady. A temperature time plot of the steel from the time it is quenched is shown to the right.

The first stage in quenching is formed to be a draw back from the point of view of effecting sharp temperature reduction. But its existence is natural in that any liquid gots vapourised and a gaseo medium sticks to an existing surface. Thus, this stage is unavoidable no doubt, but yet its time of existence can be minimized by employing certain methods aimed at inducing the second stage i.e the vapour transport stage faster. If some salt is added to the quenching medium, micro-crystals of the salt deposited on the steel surface will spurt and break up the vapour blanket. Relative motion may also be provided between the steel and the quenching medium by either stirring the steel or the liquid or both. A jet or spray of quenching medium is also used.

Hardening

A eutectoid steel (0.8% C), when cooling under equilibrium conditions consists of pearlite. The hardness of pearlite ranges from 38-45 Ro. The same steel if quenched from austenitic state and transformed into martensite, its hardness will be over Rc. 60. Thus the steel becomes harder and is said to be hardened. The process of doing so - transforming the steel into martensite is defined as hardening. Fig. 11.4 reveals the acicular structure of martensite at a higher magnification.

Fig. 11.4

Characteristics of martensitic transformation
   (i)   It is an not time-dependent athermal transformation

(ii) It proceeds with fall in temperature,

(iii) It starts at a fixed temperature (Ms) and and completes at another fixed temperature (Mf):

(iv) It is diffusionless transformation and is supposed to occur by shear;

(v)   Alloying elements effect the Ms & Mf temperatures, Ms°C = 538-316C -39Mn -19Ni- 39Cr.

(vi) Martensitic transformation is accompanies by a volume change (due to shear) which is + ve and is about 5%;

(vii) High degree of internal stresses are induced in the steel due to the martensitic transformation.

The internal stresses induced in the steel during hardening process are of complex nature. They are opposing in nature - one is due to transformation (volume increase) and the other due to temperature decrease (contraction). Quench cracks may result on this count.

Process of Hardening: The hardening process is divided into 3 stages:

(a)   Heating or Austenitisation;

(b)   Soaking; and

(c)   Quenching.

Heating of the steel is not that simple as it appears. The process should be carried out with maximum care so that scaling and decarburization are minimized, if not avoided. The temperature of heating should be selected depending upon the composition of the steel. Hypoeutectoid and eutectoid steels are austenitised above A_3 temperatures. Hyper eutectoid steels are also austenitised about the same temperatures, though complete austenitisation is not effected. On the other hand, if the steel is fully austenitised by heating to temperatures above Acm, the danger of grain coarsening is there. Hence, a certain percentage of cementite (undissolved) is preferred to more deleterious grain coarsening effect.

After the temperature of austenitisation is reached, (the furnace temperature reading), the steel should not be taken to have attained the same temperature throughout its cross section. Firstly, steels are poor conductors of heat and secondly, there is a difference between the furnace temperature and the temperature of the job in it. So, depending upon the size of the steel, it is allowed to remain at the temperature of austenitisation for a certain period of time - called soaking time. It is recommended that soaking be done roughly for 1 hour per inch of section being heated.

Quenching should be done without allowing much time to lapse between the withdrawal of the steel from the furnace and dropping it into the quenching medium or else, the steel temperature may fall to below the critical temperature. Quenching is performed for times ranging from 5 minutes to 30 minutes depending on the size of the specimen-bulky specimens need to be quenched longer to remove larger quantities of heat.

Method of quenching should be carefully controlled or else the resultant stresses may make the job warp and/or crack. Long objects should be quenched longitudinally. Irregular sections should be held in suitable blocks and/or fixtures to attain a uniform cooling.

Hardenability of Steels: As the name suggests, hardenability of a steel is the estimate of the ease with which a steel can be hardened i.e the ease with which its structure can be transformed into martensite. Various criteria contribute to an easy transformation of the steel into martensite. They are:

1. Alloying additions which shift the S curve to the right (facilitating a relatively slower cooling)
2. Alloying elements which lower the critical temperature (facilitating a lower temperature of austenitisation) and
3. Grain size.

A measure of the hardenability is the Jominy end quench test. This test uses a standard specimen of steels to be quenched under standard conditions. Even the fixture in which the specimen is placed for quenching is standardized. All these are illustrated in the Fig. 11.5. The austinitised test specimen is quickly (in less than 30 sec) placed in the fixture and the quenching water valve is opened.

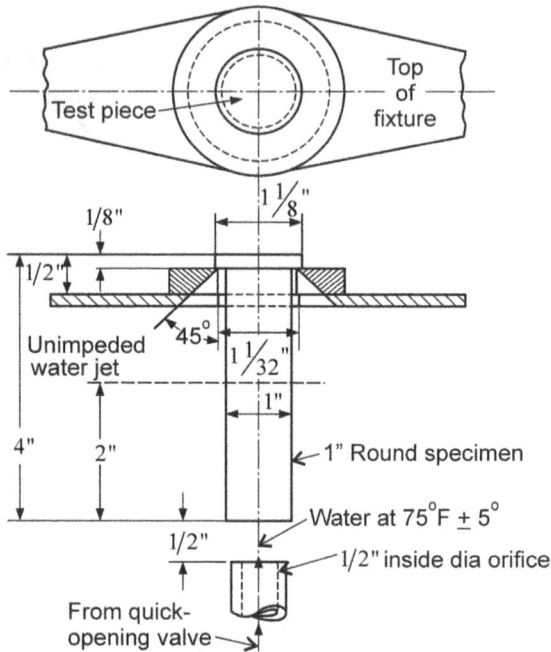

Fig. 11.5 Specimen and fixture for end-quench test.

The 1/2" water jet in pluges at the bottom surface of the specimen and quenches it. Quenching is continued till the temperature (of the specimen) reaches almost room temperatures.

The quenched specimen is withdrawn from the fixture and the flats ground on its length is again polished to clear it of any scale or dust. Hardness is measured at points 1/16" apart from the quenched and up to half the length of the specimen and 1/4" distances towards the collar. These values are tabulated.

The curve is called the hardenability curve shown in Fig. 11.6. Loads of hardenability data is available in handbooks. If the steel is having a 'good' hardinability, the curve will be reasonably horizontal with a small slope towards the right. It shows that, martensite has formed over a good distance from the quenched end. If on the other hand, the curve stoops after a small distance from the quenched end, it shows that the quench had its effect only over a small distance which hardened and the remaining length transformed to pearlite (hence has a lower hardness).

Such a steel is of a poor hardening nature steels 11.2 and 11.3 in the Fig. 11.6.

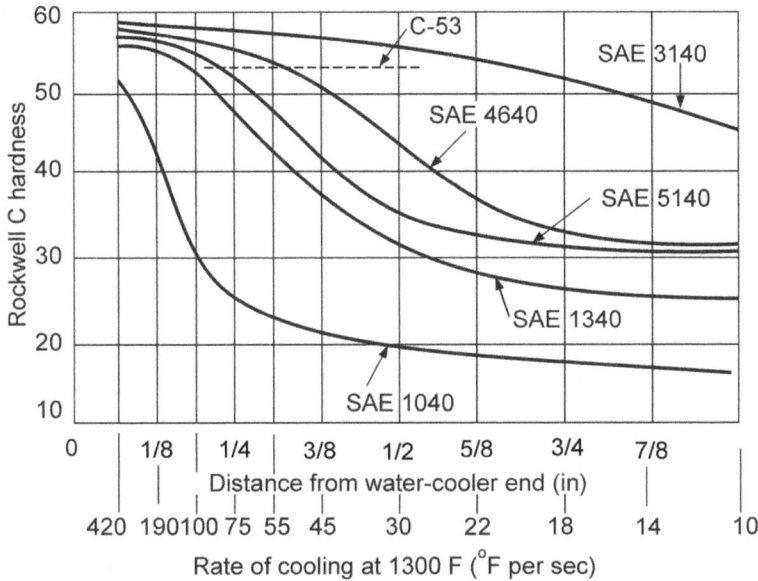

Fig. 11.6 Physical metallurgy for engine.

Ideal Critical Diameter: Ideal critical diameter (I.C.D) of a steel is defined as that diameter of the steel round whose center will consist of 50% martensite after quenching. This can also be calculated from the hardenability curve by multiplying the distance from the quenched end of the Jominy bar when 50% martensitic transformation had occurred. This distance is computed of course, from the hardness values measured.

The distance from the quenched end at which a certain hardness value is observed may be used as an indication of the hardenability. But, it should be borne in mind that the absolute value of hardness is not necessarily a matter of hardenability. The maximum hardness attainable in a fully hardened steel depends primarily on its carbon content. However it is customary to express the distance from the quenched end in a Jominy and quench test by means of a code viz. if a hardness of $-R_c$ 50 is obtained at a distance of $\frac{3}{4}''$ from the quenched and, it is written as $J_{50} = 12$.

In this relation the number 12 on the right denotes the number of the $\frac{1}{16}''$ "hardness value taken on the specimen, from the voluminous literature available on this subject, we can categorise the five factors that effect the hardenability thus,

1. Mean Composition of austenite
2. Homogeneity of the austenite
3. Grain size of the austenite
4. Non-metallic inclusions in austenite and
5. Undissolved carbides and nitrides in the austenite.

It can be seen from the above, the first the factors are inter-related. The presence of any alloying element will be felt only when it is well dissolved in the austenite. This is illustrated with the example of the alloying element chromium in Fig. 11.7. Also, the effects of non metallic inclusions, undissolved compounds etc., is to restrain the grain growth of austenite during austenitisation. The resultant five grained austenite decrease hardenability.

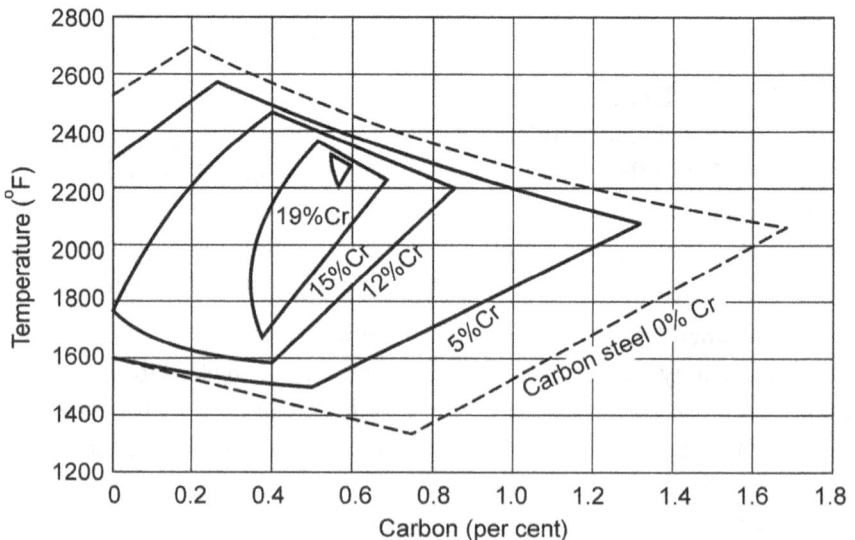

Fig. 11.7 Efffect of chromitim on the austenite phase region.
(After Bain Alloying Elements in Steel)

TEMPERING

It is a common saying that no hardened steel is used as it is. Technically speaking it is not a saying it is a rule. A hardened structure consists of primarily martensite which though hard, lacks in toughness. Thus the steel will be brittle and unusable.

A hardened steel is related to an elevated temperature depending up on the need and following the basic precautions like avoiding decarburization and the temperature not exceeding the critical temperature, the steel is held at the temperature for sufficient time when the steel

(a)   Looses some of the internal stresses developed during the hardening.

(b)

1. Some martensite breaks down to transition carbide and finally cementite.

2. Some retained austenite decomposes to bainite.

3. Carbides in alloy steels transform to alloy carbides and cementite permanently.

Tempering is done either at a low temperature below 250 °C for two to three hours. This is called low temperature tempering. Where in the internal stresses are relieved predominantly. Tempering is also same at higher temperatures up to 650-700 °C for a short time of ½ hr to ¾ hr. During this process, some realignment in the super saturated carbides will occur. Whatever may be the process the hardness of the steel falls by about 5-7 numbers and the toughness will be seen. The steel can be usable to withstand service shocks. The microstructure of a water quenched and tempered eutectoid steel in shown in Fig. 11.8.

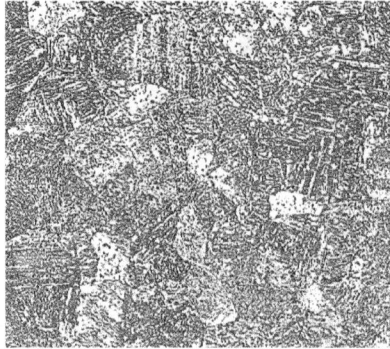

Fig. 11.8 Austine grain size revealsed by quenching to produce martensite, followed by temperating at 600 F (315 C): Velella etch (100 X)

Changes occurring the steel due to temperance, at different temperatures and the consequent properties are given below:

Tempering Temperature	Changes occuring in the steel
50–200 °C	Martensite breaks down – transition carbide ε Carbide – Fe_2^{-4} C Twins and low carbon c martensite
200–300 °C	Retained austenite decomposes to bainite. Hardness increases slightly.
250–500°C	Low carbon martensite and E carbide changes to ferrite and cementite. They precipitate along the grain boundaries. Consequently the twins coarsen gradually. The steel softens rapidly.
400–700°C	In alloy steels, carbides changes and form cementite first. The alloy diffuses in to the cementite. When it is sufficiently saturated it is substituted by another. This formation of transition carbides may be repeated many times. When equilibrium carbide forms, in chromium steels the change of carbide sequence is $Fe_3C \rightarrow C_4C_3 \rightarrow Cr_{23} C_8$. Such carbide formation sequences are very complex in steels containing more than one carbide forming alloying elements.

Temper colours: Developing of colour on the surface by the steels due to heating is a common sight in our workshops. The readers should have noticed that steel scap like springs and chips in machinery or milling with a variety of colurs. When a very thin layer of iron oxide (FeO) forms on the steel due to its micro thickness, various colours in the white light are absorbed and reflected depending on the thickness of the FeO layer. Similarly very thin FeO layers form on the steel surface when it is tempered. The steel appears in different colours depending an the thickness of the oxide layer. These are called temper colours. Some colours developed while tempering at various temperatures are given in the following Table.

Temper Colours		
Tempering Temp° C	**Colour**	**Articles tempered**
230	Pale straw	Planing and slotting Tools
240	Dark straw	Milling cutters, drills
250	Brown	Taps shear blades for metal
260	Brownish purple	Punches twist drills Reamers
270	Purple	Press tools, axes
300	Blue	Saws for wood, springs
450–650	–	Toughening of construction steels

Hardening gives a martensitic structure to the steel. This hardened steel is tempered to make the steel amenable to withstand minor shocks in source. Tempering as we have seen above relieves the internal stresses and also stabilizes some metastable constituents in the steel like alloy carbides and some cementite. Ultimately hardened and tempered steel possesses hardness in the range of 40–45 Rc and toughness in the range of 10–15 ft.lbs.

If the same mechanical properties can be attained by a single process instead of two processes of hardening and tempering, it will be a great saving not only with respect to time but also money. Similar structures and/or properties are attainable in steels by processes called austempering and martempering

Austempering: A microstructure of lower bainite or acicular bainite as it is called is developed by austempering. The steel is austenitised as is done for conventional hardening. After sufficient time of temperature homogenization (soaking) it is transferred into a both held at a temperature above the Ms to facilitate the bainitic transformation and then cooled in air or water to the room temperature. The austempering and hardening and tempering process are diagrammatically shown in Fig. 11.9.

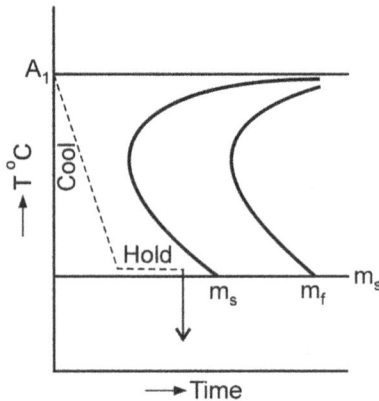

Fig. 11.9 Austempering.

The austenitisation temperature varies between 800-915° C and the austempering bath maintained at 260–400°C depending upon the steel.

We may define austempering as the process of Bo thermal transformation of a ferrous alloy at a temperature below that of pearlitic transformation and above that of martensitic transformation.

Advantages:

1. The steel will possess increased ductility, toughness and strength at a given hardness.

2. Reduced distortion means less subsequent machining time, stock removal, sorting, inspection and time.

3. Short overall time cycle to harden to 35-55 R_c with resultant time and energy savings.

The following table illustrates the significance of the austempering process;

Steel 0.74 % C

Property	Hardened & Temp at. 300°C for 30 mts	Austemped at 300° C for l5 mts
Hardness Rc.	50.2	50.4
U.T.S.P.s.i.	245.000	2, 89,000
% El. 6" g.l.	0.3	1.9
% Red.. in Area	0.7	3.4
Impact value ft. lbs	3.0	35.00

Martempering

The process of hardening and subsequent tempering can be avoided if the martensitic transformation can be brought about with minimum internal stresses in the steel. As the internal stresses develop due to the differential cooling rates existing at the surface and the core of the job, the quench temperature gradient can be minimized by interrupted quenching. The steel is first quenched in a bath held at a temperature above the Ms Temperature for sufficient time to effect temperature homogenization and finally quenched. Holding above Ms Temperature should be only for such time that it does not touch the S–curve or else fine pearlite or bainite will form. This process is also called step or interrupted quenching. This generally called martempering. It is diagrammatically illustrated in Fig. 11.10.

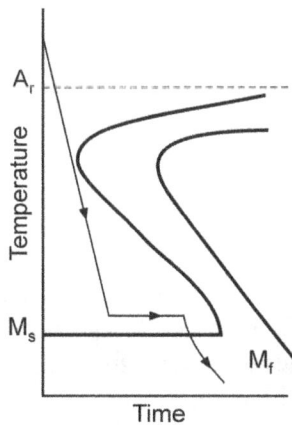

Fig. 11.10 Cooling conditions for martempering in relation to continuous cooling transformation curves.

SURFACE AND CASE HARDENING

Many engineering items require a hard surface which comes into contact with others in service. Such surfaces are required to possess good abrasion or wear resistance. It is obtained by high hardness. Often times, these parts take up loads or transmit power. E.g., Gear wheel, torsion bars, axles, shafts etc. The load distribution studies in such components indicate that the maximum stress occurs at or near the surface layers. Thus, it is enough if such components can be hardened at or near the surface, leaving the body (core) relatively soft. In fact, such products are better from the service point of view because, the tougher core can absorb shocks.

Many processes of case hardening are in vogue. All of them fall in to three main categories.

1. Those which do not involve any change in the chemical composition of the case; and

2. Those which involve a change in the chemical composition of the case.

3. Case hardening methods which do not involve any change in the composition of the case.

The steel selected for the component is of the required carbon content i.e., around eutectoid carbon (if plain carbon steel). The surface is heated by either Oxy–acetylene flame or induction current. Immediately, it is quenched; if the heating is by flame, the quenching is done by water and if it is by high frequency induction current, the quenching is done by compressed air.

Case hardening methods involving a change in the composition of the case:

1. Case carburizing

2. Cyaniding

3. Nitriding

1. **Case carburising:** Case carburizing is one of the case hardening article is increased, in order to produce a combination of wear resistance at the surface and high toughness in the core. To have high toughness in the core, it is essential to use a low carbon steel. The articles subjected to case carburizing are usually manufactured of steels of relatively low carbon content, such as 0.2 to 0.25% C. The steels ordinarily chosen for carburizing include both plain

carbon as well as alloy steels. The carburizing process should be adiusted so as to obtain a carbon content (of 0.8 to 0.9%) in the surface layer of the article treated. In any case, it should not exceed 1% as a higher carbon content of the surface layer would produce a very coarse cementite net-work at grain boundaries which would make it extremely brittle.

Process: The process of carburizing consists of increasing the carbon content of the surface by heating the article in contact with a carbonaceous medium at a temp. of 850–950° C.

In modern practice, case carburizing may be carried out with solid liquid or gaseous carburizing mixtures. Depending upon the carburizing mixture, the processes are known as "Pack Carburizing" (solid), Liquid Carburizing and Gas Carburizing.

Pack Carburizing: This is a batch process in which the parts to be carburized are packed in alloy-steel or alloy cast iron boxes along with the carburizing mixture and the boxes are dosed by luting with clay. The boxes are then heated to a temperature of about 850–950°C for a period of 4–20 hours depending upon the case depth required.

The increase in carbon content is not by direct assimilation of carbon at surface. During the heating, the following chemical reactions take place;

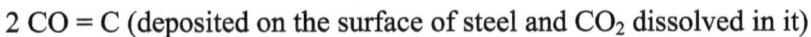

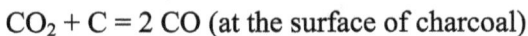

$$2C + O_2 = 2 Co \text{ (at the surface of charcoal)}$$

$2 CO = C$ (deposited on the surface of steel and CO_2 dissolved in it)

$CO_2 + C = 2 CO$ (at the surface of charcoal)

At the surface of charcoal CO forms from carbon. At the surface of steel CO dissociates into CO_2 and carbon which dissolves into the surface. The CO_2 produced again reacts with steel to form CO and the cycle is repeated many times. Thus it is clear that CO_2 acts as the carbon carrier and as such the rate of carburizing can be increased by increasing CO_2 content to a limited extent. This is the exact function of energizers to a limited extent. This is the exact function of energizers like $BaCO_3$, Na_2CO_3, $CaCO_3$ etc which dissociate at carburizing temperature to produce CO_2;

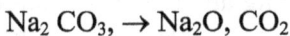

$$Na_2 CO_3, \rightarrow Na_2O, CO_2$$
$$BaCO_3, \rightarrow BaO + CO_2$$

Thus they act as catalysts.

The carburizing mixture used in pack carburizing consists of the following:

Charcoal	74-78%	Carbon Supply
$BaCO_3$	12-15%	
Na_2CO_3	1-1.5%	
$CaCO_3$	3-5%	Energizers
Fuel oil	4-5%	acts as binder or mixing agent.

Size of the charcoal-pea size (3-8 mm)

To remove dust, this should be screened through 2-3 mm sieve. The temperature most commonly employed is 850-950°C. The heating is done by placing the boxes in a furnace at a temperature of 700°C, and then slowly raising the temperature to 950°C. This time required for raising the temp, of boxes to 950°C is approximately 2 hrs per 100 mm width and height of the box.

The carburizing time is determined by the case depth required which is roughly 1 hour per 0.1 mm of case depth. In majority of applications case depths ranging from 0.5 mm to 2 mm are employed.

Liquid Carburizing: In liquid carburizing, steel is immersed in a salt bath consisting of $Ba\ Cl_2$, $Na\ Cl$, Na_2CO_3, and NaON (10-40%) at a temperature of about 925°C for a period of about 15-30 minutes. The heat treatment processes are prescribed of which the double treatment of case and core followed by low temperature tempering is a sound method. The core being of low carbon structure, needs austenitisation above A_3, Quenching gives a fine grained structure.

In the second phase, only the case portion is austenitised above A (without soaking for along) and quenched so that it is transformed into Martensite. Finally the steel is low temperature tempered.

2. **Cyaniding:** In cyaniding, both carbon and nitrogen are allowed to penetrate the steel. The baths consist of sodium cyanide, sodium carbonate and sodium chloride. In high temperature cyaniding Barium chloride is also added. The temperature of cyaniding vary from 550° C to 950° C, depending upon the particular requirement. The case depths obtained vary from 0.04 mm (low temp. process) to 1.5 mm (in high temp. cyaniding). The times of cyaniding also vary depending upon the temperature of the process – 30 mts in low temp. to 4 hrs, in high temperature process. In low temperature

cyaniding, nitrogen diffusion is found to be predominant and in the high temperature cyaniding, carbon diffusion is predominant, leading to the necessity of subsequent heat treatment.

Parts commonly cyanided are the power transmission chain drums, small pump gears, worm drive screws barke cams etc.

3. **Nitriding:** Nitriding process consists of subjecting meachined and heat treated steel to the action of a nitrogenous medium commonly ammonia gas, thereby imparting a high surface hardness. The steel is machined and heat treated for the required properties before nitriding.

The process is a simple gas exposure process, the steel is maintained in a furnace chamber at a temperature of 500 °C - 600°C and ammonia gas is allowed into the furnace chamber. The time of nitirding ranges from 24 hrs to 72 hrs depending upon the steel being nitride.

Nitrogen gas is formed inside the furnace due to the decomposition of ammonia $2NH_3 \rightarrow 2N + 3H_2$. Partially, the nascent nitrogen combines with the alloying elements present in the steel to form nitrides, The nitrides, in a fine state of dispersion in the case, impart extreme hardness to the surface of the steel.

Precaution: Cyanide is a deadly poison. All workmen should wear hand gloves. Even the cyanide fumes are deadly and workmen near the cyaniding pots should wear masks. The cyanide contents in the industrial effluents should be neutralized before they are let out.

CHAPTER 12

Cast Irons

We have defined all the iron carbon alloys containing carbon in excess of 2.00% as cast irons. Thus cast irons are the iron carbon alloys which contain carbon in the composition range 2.00–4.5 percent. Here it is better to understand the common term "iron" used to refer to cast iron. The word iron should be understood in the context used but not misunderstood as pure iron.

As we know that the solubility of carbon in iron at room temperature is 0.01 percent and carbon can exist as pearlite up to 0.8 percent, the quantity existing in excess of 0.8 percent will have to be either free carbon or in the combined form of cementite. Cementite usually gets well distributed in the structure and makes the item extremely hard and brittle Seen under the microscopes it looks brilliant white and hence the irons containing such structure are known as white cast irons. Photographs of irons containing different microstructures are given for easy understanding. These photographs are called photomicrographs. Certain conditions of cooling and alloy additions are necessary to obtain a white cast iron casting. Cementite Stabilizers i.e. carbide formers are added to the iron and it is generally fast cooled. Fast cooling does not permit graphite formation and casting remains white cast iron. A fractured surface looks white and hence the name.

Under normal casting practice, the cast iron cools with the separation of a small quantity of ferrite, some peralite and graphite. The graphite exists as flakes. The area of graphite flakes acts as these raiser area and the iron has a low impact strength. Its fracture appears grey due to the free carbon (in the form of graphite flakes). These irons are called grey cast irons, metallographically, they are also called flake graphite cast irons. Cast Irons commonly used fall into this category only (see Fig. 12.1).

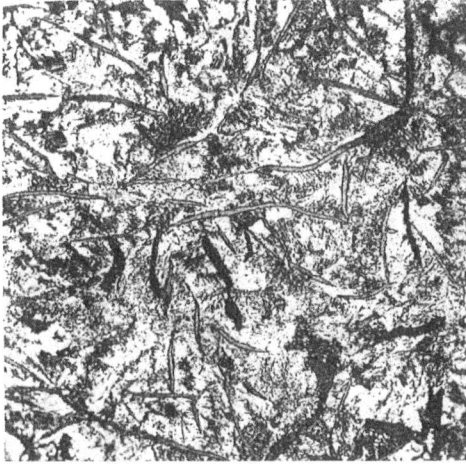

Fig. 12.1 Gray case iron (250 X).

The existence of carbon in flake form was found to be the cause for the relative weakness of cast iron. The flake size was studied and it was proved that irons with a fine graphite flake distribution were found to be stronger than those with coarse graphite. On strength basis irons are classified in to three varieties – Light duty cast irons which possess tensile strengths up to 2000 Kg/cm². Irons whose strength ranges from 2000 to 3500 Kg /cm² are classed as medium duty cast irons and those possessing strengths above 3500 Kg/cm² are called heavy duty cast irons.

The percentage elongation symbolizing ductility is very low in cast irons – a maximum of 5%. Compression strength of cast irons very high. It is about 2.5 to 4.5 times its tensile strength. Hence it finds use in machine bodies and bases where load bearing is the only strength criterion. The hardness of cast iron ranges from 80-200 B.H.N. But the toughness measured as the Izod impact strength is very low – a mere 10 ft. lbs. This is as expected due to the existence of graphite flakes, each of which is a stress-raiser.

When an iron casting is chilled – cooled very fast, its surface and same depths solidify with a white iron texture and the central portion where the effect of chilling will not be felt, will solidify in the usual grey iron form. Items which need to process a high wear and abrasion resistance and yet possess good toughness line ball – mill balls are made in this process. Such a cast iron is named as chilled cast iron or mott led iron. (See Fig. 12.2).

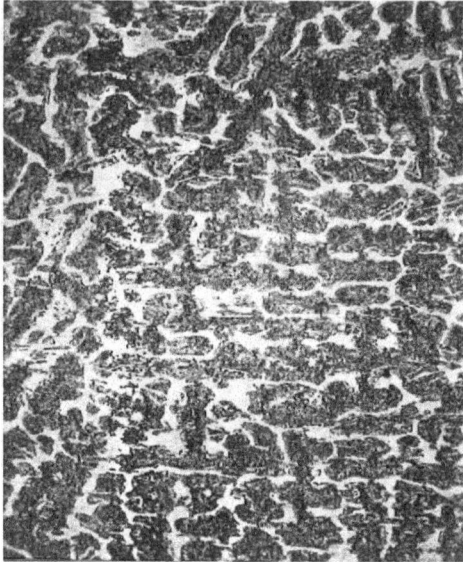

Fig. 12.2 White case iron (200 X).

Composition of cast iron: The usual alloying elements present in iron are carbon, silicon, Manganese, sulphur and phosphorus besides some trace elements.

Carbon: Carbon lowers the melting point. It also produces more graphite. Hence a soft weak structure results.

Silicon: Silicon strengthens ferrite by virtue of dissolving in iron. So it indirectly promotes the breaking up of cementite. When the cementite breaks up, carbon is liberated and graphite is formed. That is why silicon is known as a graphitiser. Maurer diagram shown in Fig. 12.3 gives the relation between the carbon and silicon contents in producing different irons for one rate of cooling. It can be seen that either high carbon and silicon or low carbon and high silicon percentages produce grey iron.

Manganese: Manganese when added to iron reacts with the iron sulphide present and forms manganese sulphide. Iron sulphide thus broken up the tendency to form cementite is minimized. So manganese indirectly promotes the formation of graphite. Further, when it exists in larger quantities than that just required to react with the iron suiphide, it hardens the iron. Roughly 1.7 parts of manganese is required to react with one part of sulphur.

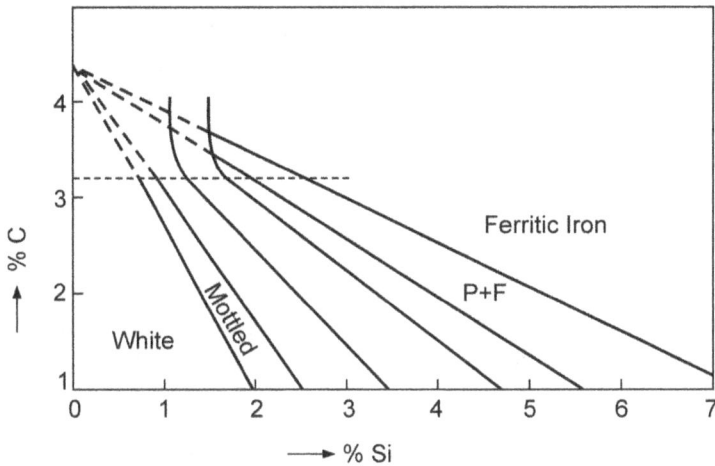

Fig. 12.3 Maurer diagram.

Sulphur: This is in fact not an alloying element. It is classed as an inevitable impurity finding its way in to the metal from coke during iron production in blast furnace. Its deleterious effect is neutralized by manganese and sometimes by calcium.

Phosphorus: This element has no effect on the graphite or cementite formation processes. Phosphorous forms an iron phosphide eutectic which is low δ melting. So the iron becomes very fluid and gets segregated at the grain boundaries, leading to a weak casting. That is why phosphorus should be kept to a minimum is the production of large iron castings in sand. A phosphorous content not exceeding 0.3 per cent is recommended.

Trace elements: Certain elements which are not normally considered to be alloying elements in general have remarkable effect on the characteristics of cast iron. 0.1% of aluminimum acts a good graphitiser while a microscopic quantity, 0.003% of hydrogen reduces the strength of iron by coarsening the graphite. Lead and tellurium are found to yield unsound castings with porosity etc, even in small percentages (say 0.1%). Nitrogen behaves as a carbide stabilizer.

Carbon Equivalent: The eutectic composition of 4.3% C, is naturally the lowest solidifying temperature of cast irons. It is found that the irons with more carbon (under some conditions) give out large sized graphite during solidification. This graphite is called Kish graphite. Carbon contents of less than 4.3% yield normal graphite structure except when a mottled or white iron composition range is reached. Some elements, especially silicon and phosphorous affect the eutectic composition in an alloy consisting of more alloying elements. It is suggested to calculate what is called 'carbon equivalent' value to exactly understand their combined effect, instead of a carbon percentage only. The carbon equivalent value guides us to evaluate how close the composition of the iron is to the eutectic. Then we can predict successfully how much free graphite will form in the iron on solidification.

$$\text{Carbon Equivalent} = \text{Total C} + 1/3 \ (\% \ S + \% \ P)$$

Other elements: Nickel is a graphitizer. It not only controls graphitization but also tends to control the size of the graphite flake. In its graphitizing capacity, it can be assumed to be half as powerful as silicon. Addition of judicious amount of nickel, the strength of the iron is increased. Generally, nickel is used up to 4%. However an alloy containing 4.5% Ni, 1.5% Cr, high carbon and low silicon is known by the commercial name Ni-hard possesses an outstanding wear resistance. Another alloy with 15% Ni, 2% Cr, and 6% Cu besides normal other elements is known as Ni-resist. This alloy is austenitic and possesses very good strength growth and high temperature resistance up to 800° C. Chromium is added up to 0.5% to reduce growth of the iron.

Molybdenum imparts very beneficial properties to the iron and is added in small quantities up to 1.2%. Molybdenum is not a graphitiser but a mild carbide former. It dissolves in ferrite thereby strengthening the matrix and controls better graphite distribution.

Vanadium is a strong carbide former and is added in small quantities (0.25%). The action of vanadium is some what similar to molybdenum as far as strengthening the iron.

Copper is used to strengthen low carbon cast irons. It is a mild graphitiser.

Growth of cast irons: When liquid cast iron solidifies and graphite forms in the solidified iron, that much weight as a percentage of the total casting weight will increase in volume. Graphite has a density of about 1 while the density of the alloy from which it had separated is nearly seven times that value. So, the volume increases. The dimensions of the casting are found to be 2-3% more than the mould made, depending upon the percentage of carbon in the cast iron. This increase in dimensions is known as 'growth' of iron. Due care is taken by the foundry man while making the cast iron casting to overcome this growth.

White Cast Iron: Iron carbon alloys containing more than 1.7 per cent carbon in the form of iron carbide are called white cast irons. They behave in accordance with the equilibrium diagram of iron carbide. The portion of the equilibrium diagram between 1.7- 4.3 are hypo eutectic cast irons and those in the carbon percentage range 4.3 - 6.67 are hyper eutectic cast irons. Leduberite eutectic composition is 4.3% C. Microstructures of a hypo eutectic and eutectic white cast irons are shown in Fig. 12.4 and Fig. 12.5.

Fig. 12.4 Hypo-eutectic white case iron, cementite and pearlite (black) (X 100) B.J = 500.

Fig. 12.5 Hypo-eutectic white case iron. X 100. White primary crystals of cementite in eutectic (cementite and pearlite)

White cast iron is not of much structural use owing to its excessive brittleness. Its mechanical properties vary over a wide range. Their tensile strength varies from 2500 kg/cm^2 to 5000 kg/cm^2 and hardness increase as the percentage of carbon increases (because more of cementite is formed). The tensile strength of white cast iron usually encountered will be about 3500 kgs/cm^2 and hardness 350-450 BHN. Of course, its compression strength is around 40,000 kgs/cm^2. These cast irons are of limited use for items requiring wear and abrasion resistance like wearing plates, parts of abrasions machinery, sand slingers etc.

Malleable Cast Iron: The uniqueness of malleable cast iron lies in the fact that graphite is manipulated by a suitable heat treatment in the solid states only. The combined carbon existing as cementite is made to diffuse out and segregate at suitable sites as lumps. These are called 'rosettes' of carbon. Carbon, thus existing in nodular shape instead of flakes dramatically improves the properties of the iron (see Fig. 12.6). Cast Iron objects are made in white cast iron structure with suitable alloy manipulation to avoid graphitization during cooling from the molten state. All the carbon exists in the combined form as cementite. The process of malleablising starts now. The white cast iron castings are packed in boxes with a mixture of already once used hematite ore and new hematite ore and heated to a temperature over 900 °C and maintained at that temperature for four to five days. With gaseous fuel heating, the time of annealing will be reduced to 3 days. The structure of iron changes to one similar to steels, containing pearlite and ferrite interspersed with nodules of graphite. The fracture of the iron, after malleabilising appears white and so the iron is called White heart malleable iron.

White cast iron castings are also annealed in a neutral atmosphere at 900°C. Due to the control in composition of the castings in regard to the silicon and sulphur percentage, cementite breaks up into ferrite and graphite. The malleablising cycle can de divided into three steps. The first step consists of heating the castings very slowly at a rate of 20°C per hour up to the malleablising temperature. Slow heating favours the formation graphite nuclei. The second step consists of soaking for about 40 hours for the growth of the nuclei. The third step is slow cooling to allow carbon precipitation from austenite. This prevents the formation of cementite again. This is followed by very slow cooling through the eutectoid temperature range (725-670°C). The iron developed is known as black heart malleable iron due to fracture appearing grey at the centre.

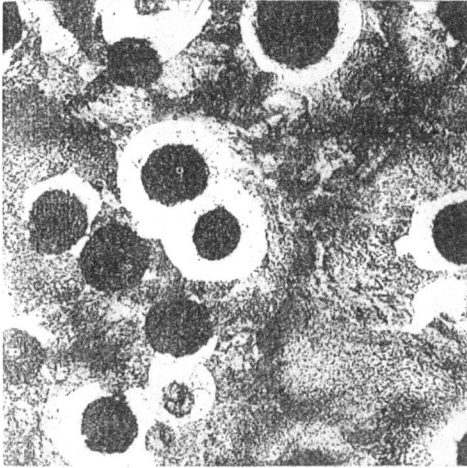

Fig. 12.6 Nodular case iron as cast (250 X).

Another type of malleable iron is also produced which consists of a pearlite matrix and is known by the name peralite malleable iron. The malleablising process is similar to the black-heart process except of the second step i.e., instead of very slow cooling which was done to avoid pearlite formation – the iron is more rapidly cooled and in some cases an alloy addition of 1% Mn is also resorted to. The matrix would be of pearlite. This iron has excellent wear resistance properties. It is used in the manufacture of gears, links etc.

Mechanical Properties of Malleable Cast Irons

Property	White heart Malleable		Black Heart Malleable		
	Gr I	Gr II	Gr. I	Gr. II	Gr. III
Tensile Strength	2000 Kg/Cm^2	2500 Kg/Cm^2	1800 kg/Cm^2	2000 Kg/Cm^2	2200 Kg/Cm^2
Yield Point	1000 Kgf	1200 Kgf	1200 Kgf	1200 Kgf	1300 Kgf
% Elongation	5	10	5	10	14

Spheroidal Graphite Cast Iron: It can be seen that the superior properties of malleable cast iron are due to the coagulation of graphite as rosettes. This has avoided the stress raising graphite flake structure. Subsequently it was discovered that an iron which solidifies as a grey cast iron can be made to solidify with a structure consisting of graphite in spheroids. For this to happen, certain trace elements like cerium and magnesium are added to the molten metal in the mould. Spheroidal graphite cast iron name is given to the iron, such an iron possess properties comparable to those of malleable cast iron. A great advantage in this is the costly and laborious processes of malleablising is done away with. The microstructure of a spheroidal graphite cast iron is shown in Fig. 12.7. The mechanical properties as said above, vary from 2000 to 3000 kg /sq cm strength with the percentage elongation varying from 15 to 5 respectively. The higher strength ranges have a pearlite matrix.

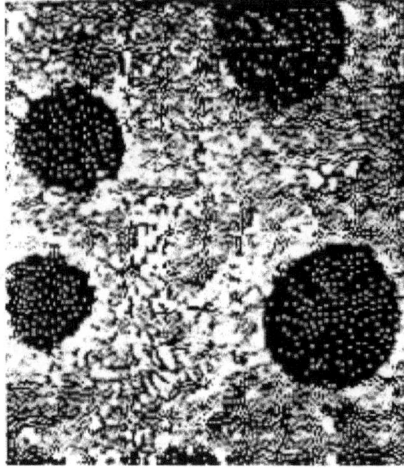

Fig. 12.7 Spheroidal cast iron, spheroidal graphite in pearlite matrix, X 200.

Alloy Cast Irons

The effects of different alloying elements on the microstructure is explained above. The properties of individual phases broadly known like ferrite soft, cementite hard and so on, the alloy composition in a particular iron can be manipulated depending upon the properties required of the iron. However some typical alloy iron compositions are discussed below.

Wear Resistant irons: These are hypoeutectic irons containing chromium and molybdenum each ranging from 0.6 to 1.0 percent. Because of the fine dispersion of alloy carbides in the microstructure, the iron exhibits good wear resisitance. Such irons find application in engine liners, press sleeves dies and such parts.

Acicular Irons: This cast iron derives its name from the microstructure. It contains ferrite in acicular form (like needles) in an austenite matrix. Such a structure is obtained by suppressing the eutectoid point by judicious alloy additions of nickel and molybdenum. Sometimes copper is also added instead of nickel.

The acicular cast irons possess high mechanical properties. They exhibit good toughness. Irons with 3.00% carbon have up to 3200 kg/cm^2 tensile strength. However they cannot be used for service at temperatures above 300°C. It is important to keep in mind the harmful effects of chromium and phosphorous. Chromium is kept within the 0.4% level whereas phosphorous should not exceed 0.15%.

Martensitic Cast Iron: This is also known by the service name Ni-hard. It consists of a martensitic matrix with grains of cementite. They have very high abrasion resistance and generally used in making rolling rolls.

Austenitic Irons: These are non magnetic, containing mainly nickel up to 15 percent. Details of Ni-hard and Ni-resist are given in the preceding pages

Note: During the micro-examination of irons, the polished specimen can be viewed under the microscope directly. Graphite present in the structure will be revealed without any etching. However, other constituents will only be revealed by etching with a suitable etchant.

COPPER ALLOYS

Pure copper is essentially used for electrical conductors. Oxygen, if present even in traces is found to decrease the electrical conductivity of copper dramatically so, copper is cast into electrode shaped ingots and refined electrically. This copper is used for electrical cables and is known by the metallurgical name oxygen free high conductivity copper (OFHC). Apart from the electrical usage, Copper is used in alloy form. Alloying increases the strength of copper.

Common metals that are alloyed with copper are zinc (in brasses), tin (in bronzes) and others like nickel, aluminum, silicon, calcium and beryllium. Study of copper alloys is rather confusing because of the haphazard way in which the names are given to the alloys. Many brasses as they should be known are in fact called bronzes. An alloy called nickel silver does not contain any silver but copper, nickel and zinc; silvery appearance may be the reason.

Copper Zinc Equilibrium: The copper-zinc equilibrium is shown in Fig. 12.8. It can be seen that more than a thing the diagram is occupied by α solid solution. Further nearly 50% of the composition range is covered by the two solid solutions α and β. The alloys containing zinc up to 38% are called α brasses and above that α, β brasses. The mechanical properties of brasses are very closely related to the equilibrium diagram. The tensile strength and ductility increase as the percentage of zinc increase in a brass, up to 30% Zn. Thereafter, with the appearances of β solid solution, the ductility starts to decrease though the strength increases. The solid solutions, β and y are not so ductile as α, and so no brass containing more than 40% Zn is in practical use.

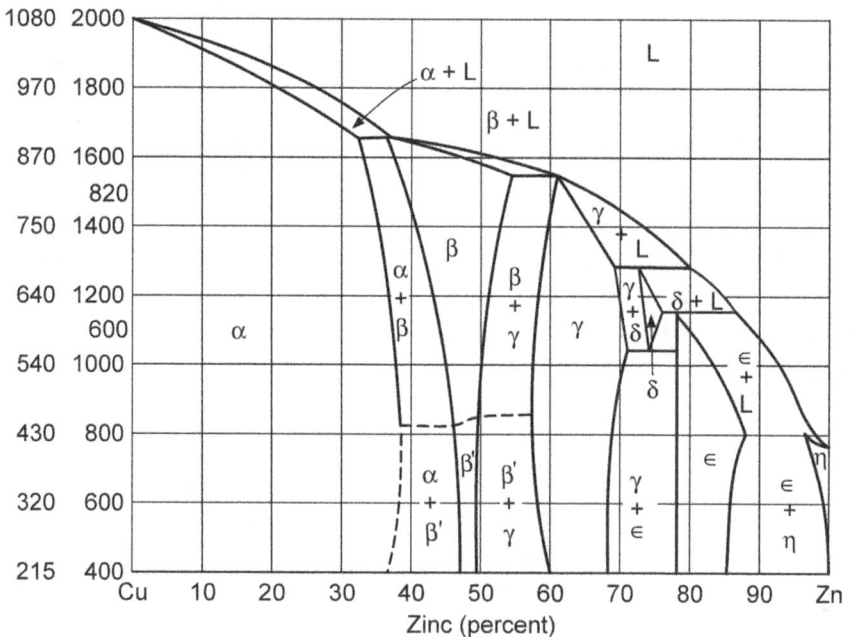

Fig. 12.8 Equilibrium diagram for copper-zinc system.

Copper-tin equilibrium: Alloys which principally containing copper and tin are considered as bronzes. These alloys posses good strength, wear resistance and salt water corrosion resistance. An equilibrium diagram of copper and tin is shown in Fig. 12.9. Unlike copper-zinc system, this system has homogeneous alpha phase up to 16% S_n. But, the solubility of tin at room temperature is restricted. With more quantity of tin, a hard compound Cu_3Sn, epsilon appears which restricts ductility. Many bronzes however, are not pure copper, tin alloys as some desirables additions of other elements are always made. These additions make bronzes possess desirable characteristics. Thus, it can be said that the useful composition range is up to 20% tin. The entire bronze family can be made into four groups as here under,

1. Bronzes containing 0-8% tin. These bronzes are used in the making of plates, sheets, wire and coins.

2. Bronzes containing 8-12% tin. These bronzes are used for the manufacture of machine parts, especially gears, bearings and machine fittings.

3. Bronzes containing 12- 20% tin: these bronzes are especially used for the manufacture of bearings. These are also called bearing bronzes.

4. Bronzes containing 20 - 25% tin: These are also called bell metals. Used in the manufacture of bells. The items are mostly cast.

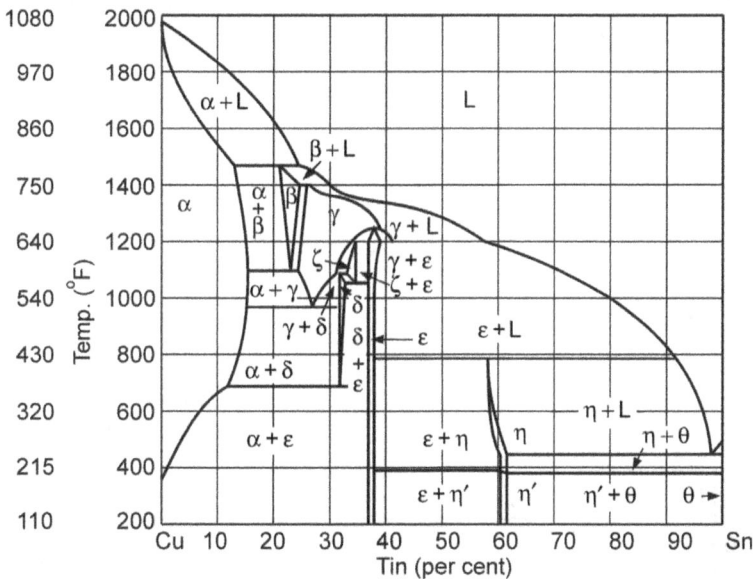

Fig. 12.9 Equilibrium diagram for copper-zinc system.

It will not be out of place to mention about Phosphor bronzes. Phosphor bronzes contain tin, generally in the range mentioned in the first group above. Generally no other alloying elements are used. Phosphorus is added for the purpose of de oxidation. Without copper or tin oxide in the structure, the strength and ductility of the bronze is improved. Being a better hardener than tin, the addition of phosphorus leads to the formation of Cu_3P, which is hard and improves the general soundness of the casting. Usually, phosphor bronzes contain about 0.1 - 0.3% P and 10% Sn. These alloys are known for their very low co-efficient of friction. With a tensile strength of 2500 kg/cm^2 and elongation of 5% and BHN 100, they are commonly used in the manufacture of delicate gear wheels, slide valves and springs. Very good corrosions resistance makes phosphor bronze an automatic choice as the material for turbine blades also.

Gun metal: It was used to make the barrels of the guns in the olden days and hence the name. A bronzes of the very common 88/10/2 type-(88 Cu, 10 Sn, 2 Zn), it has a good tensile strength and elongation. It is a hot working alloy and possesses a good wear and corrosion resistance.

Copper Nickel system: Copper and Nickel are soluble in both liquid and solid state and form a primary type solid solution α at room temperature. An equilibrium diagram is shown in Fig. 12.10. This equilibrium is similar to the type I equilibrium system studied under equilibrium diagrams earlier.

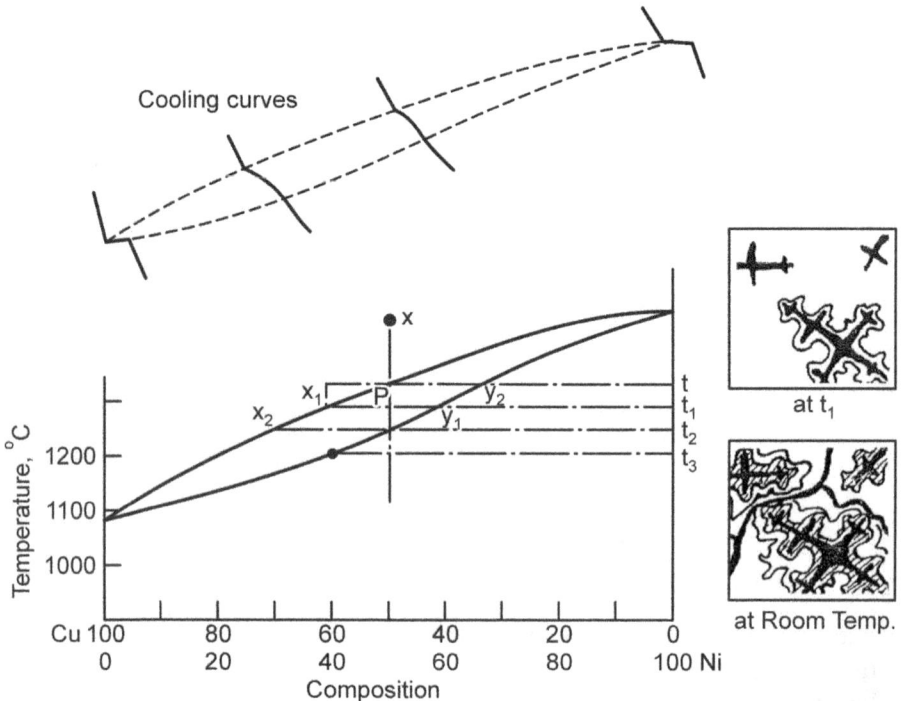

Fig. 12.10

A few alloys in copper nickel system are of specific interest. They are commonly known as Cupro-nickels. These alloys contain varying and 5% zinc is known by the name nickel silver is widely used for coins. Another alloy with 55% Cu, and 45% Ni is known by the name Constantin, is a thermocouple element.

A family of alloys with trade name monels are highly corrosion resistant and wear resistant. They possess strength and hardness in the range of mild steel. One type of monel called K-monel is heat-treatable. It can be quenched from 800°C and tempered at 600°C for 4 hours. It is quite strong and ductile. It is used to manufacture turbine blades owing to its corrosion resistance to sea water and alternative solution as well.

Copper aluminum system: The copper end of the equilibrium diagram of copper aluminum system is shown in Fig. 12.11. The alloys commonly

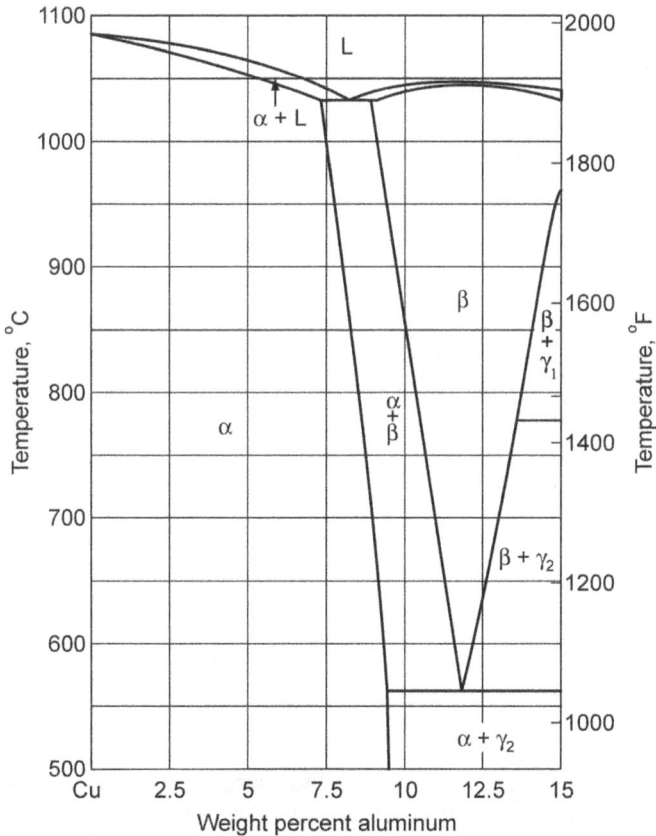

Fig. 12.11 Copper-rich portion of the copper-aluminium alloy system.
(From "Metals Handbook,"

known as 'aluminum bronzes' usually contain 10-11% aluminum and 4% iron, and only very small amounts of tin. Sometimes, nickel up to 1% is also added. These alloys are very hard and difficult to machine. These to phase's alloys have good strength, wear resistance, corrosion resistance and also possess good fatigue and impact strengths. Heat treatment consists of quenching from around 850°C and tempering at 400-550°C, to develop optimum properties.

Many castings of aluminum bronzes are also in engineering use. These casting need careful foundry practice owing to their susceptibility for shrinkage.

Uses

    (a)   Gears in heavy machinery

    (b)   Feed nuts

    (c)   Bearings and valve guides

    (d)   Non-sparking tools

    (e)   Cold working dies

    (f)   Electrical contacts

Copper-silicon system: Fig. 12.12 shows the copper end of the copper silicon equilibrium diagram. It can be noticed that silicon is soluble in copper up to 5.3 percent at 845°C. The solidus curves slants towards

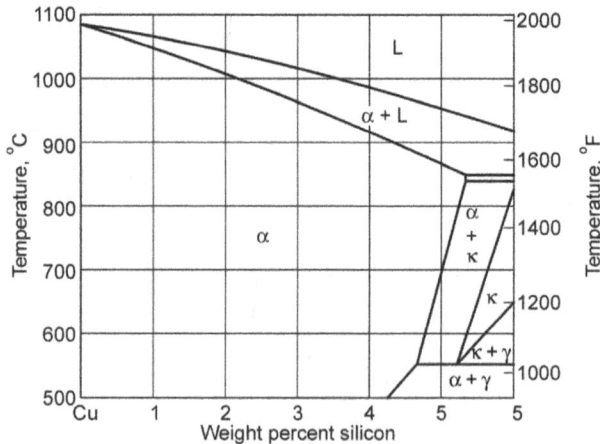

Fig. 12.12 Copper-rich portion of the copper-silicon alloy system
(From "Metals Handbook," 1948 ed., P. 1203 American sociery for metals park, Ohio).

copper indicating that the solubility decreases with temperature. Thus at room temperature, the solubility is only about 3.7% Si. Majority of these alloys are commonly called silicon bronzes.

These alloys owe their industrial importance due to their excellent mechanical properties and corrosion resistance.

Copper Beryllium alloys: Beryllium is soluble in copper up to 2.1% at 870°C and the solubility decreases with temperature. At room temperature, the solubility is around 0.25%. So, the alloys are age-hardenable.

Beryllium metal is extremely costly and so it is kept in the range of 1.5-2.2 the alloys. These alloys are amenable to the cold working also. By a combination of the two methods viz., cold working and age hardening, tensile strengths up to 20,000 kg/eut and B.H.N 400 are achieved. However in such a state, the alloy records a low ductility only 2% on a 50 mm.g.l). Because of high cost, copper Beryllium alloys are used only when cheaper alloys cannot fulfill the required service requirements.

Aluminum: This metal is the second most important and predominant among non- ferrous metals. Aluminum of 99 percent purity is very soft, T.S. 1200 kg/cm^2 with 40% elongation. Consequently, it can be successfully cold-worked and in work-hardened condition, its strength will be double that of the strength in annealed state.

Aluminum has a natural corrosion resistance property. Being highly reactive, it readily gets oxidized and forms $Al_2 O_3$. $Al_2 O_3$ is a very stable and corrosion resistant oxide. In fact, it is so non reactive and impervious that $Al_2 O_3$ of a few atoms thick on aluminum prevents further formation aluminum trioxide by insulating the metal from atmospheric oxygen. Many aluminum alloys are made corrosion resistant by taking advantage of this phenomenon. Al_2O_3 is formed on the surface of the object. This process is called anodizing.

Aluminum alloys are also made stronger by cold work, as mentioned above or by a process called precipitation or age hardening. The most important in this category are those alloys belonging to the 'duraluminium' family. These are mainly aluminum copper alloys with traces of other elements like iron, manganese etc.

Age hardening: Age hardening or precipitation hardening was discovered by A. Wilm in 1906. He noted that an aluminum alloy containing copper which was quacked from a relatively high temperature developed a higher hardness on keeping at room temperature for some days. Because of the time in days involved the termed it as 'age-hardening'. This phenomenon is also found to occur with numerous alloys, both ferrous and non-ferrous. An important criteria for age hardening to occur is that the solidus line in equilibrium diagram should bend towards left-indicating a decreasing solid solubility with temperature. An alloy containing 4% Cu is annealed for sufficient time at 500°C. All the Cu Al_2 goes into the α-solid solution, (Microstructure X_1). If the alloy is slowly cooled, Cu Al_2 is precipitated 1 as relatively coarse particles which can be noticed under an optical microscope. On the other hand when the alloy is quenched from 500°C, the supersaturated solid solution is retained at room temperature (shown in structure above).This phase is slightly hard but more ductile than the slowly cooled sample. The slowly cooled sample consists of a relatively coarse Cu Al_2 particles. This however can be accelerated by heating the alloy to a slightly elevated temperature, say 200°C.

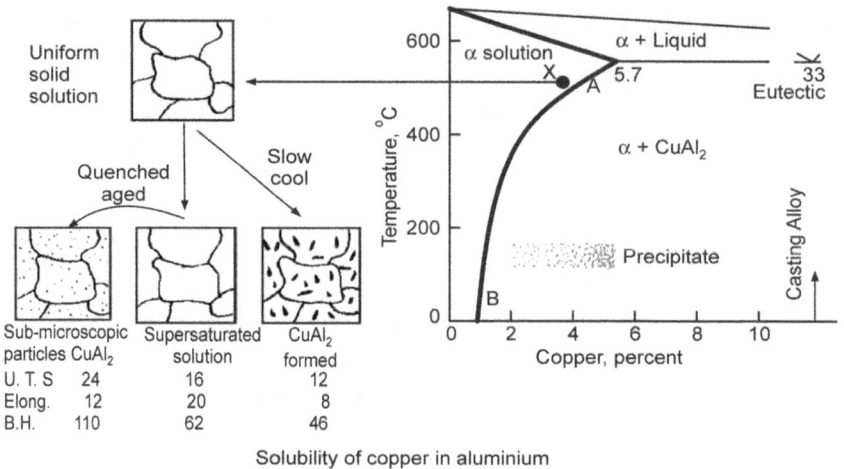

Solubility of copper in aluminium

Fig. 12.13 Solubility of copper in aluminium.

The actual reason for age hardening is not definitely known. A general explanation that the dispersion of fine particles of Cu Al_2 obstructing the slip is the reason for the hardening does not hold good for the maximum hardness occurs before any precipitate is detected in structure.

Mechanism of age hardening: It is said that age hardening takes in two stages. The first step is the precipitation from the supersaturated solid solution is the segregation of copper atoms into clusters or platelets of a few angstroms in thickness and hundred angstroms in diameter but still part and parcel of the Lattice. These are often called Guinier-Preston zones I or simply G.P.I; zones to form larger ones called G.P.II zones or Q". These are 8 angstroms thick and about 150 angstroms in diameter.

Q" has a tetragonal structure which fits into the aluminum unit cell perfectly in two directions but not in the third direction. Aluminium plates have to be distorted in order that the matrix and precipitate match each other or be coherent. This gives rise to coherency strains. The strain fields have a much larger effective size than the precipitate, oppose the movement of dislocations i e, harden the alloy.

At a still later stage a new transition phase Q' which is partially coherent with the matrix is precipitated imparting maximum hardness to the alloy. Over aging however converts this transition precipitate into the equilibrium precipitate Q or Cu Al$_2$. Thus, a structure, similar to the one obtained by slow cooling is obtained and associated with it the poor properties.

Bismuth- cadmium equilibrium: This a simple eutectic system with not even traces of mutual solid solubility. The eutectic mixture consists of Bi and Cd at 40% cadmium. While cooling an alloy of 40%Cd. When the liquid temperature reaches the eutectic temperature (143 deg C) three phases are in equilibrium. They are liquid, 60% Bi, and 40% cadmium. At this, there are no degrees of freedom and system is univariant. When one of the phases is eliminated i.e., the liquid, by cooling the remaining two phases, 40% Cd and 60% bismuth, coexist over a range of temperature.

It should be understood from the above diagram that all the alloys in the composition range 0-40% cadmium solidify as two phase alloys consisting of Bismuth and eutectic mixture Bi Cd and those above 40% Cadmium as eutectic mixture and cadmium. Of course an alloy of 40% cadmium will be a single phase alloy of Bi Cd eutectic mixture.

Copper silver equilibrium

This system is a simple eutectic system, the eutectic occurring at 72% Ag. The two metals are partially soluble in each other forming terminal solid solutions on both ends of the equilibrium diagram. The slope of the solidus lines indicates that there is scope for age hardening. An important alloy in this system is sterling silver which contains 7.5% copper. Another alloy with 10% copper is also notable. Both these alloys are used for minting coins. Other important uses are for making electrical contacts. The eutectic alloy is used are brazing solder in the manufacture of jewelery.

CHAPTER 13

Introduction to Composite Materials

Composite Materials are new kind of materials which have been used in all fields of engineering and non-engineering, such as, Aerospace, Civil, Mechanical, Marine, Robotics, Electrical, Electronics, Chemical, Medical, Bio-technology, Nano-technology, Agriculture, Horticulture etc. Composites have been used for different kinds of applications ranging from structural to aesthetic.

It is a fact that technological development depends on the advances in the field of materials. The material research is an important parameter for the development of sophisticated technology. The most advanced turbine or aircraft design is of no use if adequate materials to bear the service loads and conditions are not available. Whatever the field may be, the final limitation on advancement depends on materials. Composite materials in this regard represent a gigantic step in the optimization in materials. The idea of composite materials is not recent one. The nature is full of examples wherein the idea of composite materials is used.

NATURAL COMPOSITE MATERIALS

Nature itself is a full example of natural composite materials ranging from animal body composition to a isotropic material at the crystal level (micro level) then to Nano level and further. Here, I will brief about some of the natural composite materials, which are as follows.

1. Coconut palm leaf is an example of natural composite material. It is a cantilever using the concept of fiber reinforcement.

2. Wood is a natural fibrous composite material. The constituents are cellulose fibres in a lignin matrix. Cellulose fibers have high tensile strength but are very brittle (i.e., low stiffness), while lignin matrix combined with fibers helps in obtaining more stiffness.

3. Bone itself is a natural composite material. The bone consists of a short and soft collogen fibers embedded in a mineral matrix called

apatite. In addition, Bones are like fibers and muscle acts like matrix material, therefore the whole body itself is a composite material.

These are some examples of natural composites, similarly we can find many examples, which are not explained in this text book. In addition to these natural materials, some engineering composite materials were also in use for long time, with or without their knowledge.

ENGINEERING COMPOSITE MATERIALS

Some engineering composite materials were in use for long time, in addition to the natural occurring materials, which are

1. **Carbon block in rubber:** Carbon block in rubber is a composite material. Carbon blocks were used as particles in a rubber material to enhance the strength and stiffness of the composite material.

2. **Cement or Asphalt mixed with sand:** Cement or asphalt mixed with sand is a composite material. Sand as particles when mixed with cement or asphalt gives very good strength and stiffness and it is a common practice that everywhere it is used.

3. **Glass fibre in Resin:** Glass fibres mixed with resin is a composite material. Fibres are available in different forms either long or short, each will have different applications. Long fibers are used for laminated composite constructions whereas short fibers are used for short fiber composite constructions. These fibers when mixed with matrix materials provide enormous amount of strength and stiffness.

These are some examples of natural and engineering composite materials, now let us go back to the history of the composite materials, to know more about the composite materials and its evolution.

HISTORY OF COMPOSITE MATERIALS

Ancient people were using composite materials for their day-to-day activities without the knowledge of the concept of the composite materials. There is a long history of usage but their beginnings are not known and records are not available for study. Let us see how people were combining different materials for getting best material properties from their constituents. Isralites were using straw to strengthen mud bricks. In olden days, Ply wood was used to achieve superior strength, resistance to thermal expansion, resistance to swelling etc. Medieval swords and armour were fabricated with layers different materials. Bamboo wood has been used to build houses, bridges etc., as a structural material to carry tension, compression and shear loadings. Bamboos provide very good strength and stiffness to the structure, hence these have been used extensively for many structural members even today. These are some of the examples, where composites have been used by ancient people and there may be many more such examples and are continuously still used in many parts of the world.

For long periods, engineering and non-engineering community, were using composite materials for structural and non-structural components, without much deeper understanding about composite materials. After the invention of Aeroplanes (invented by write brothers) and space crafts, researchers started looking for lighter, stronger and stiffer materials. During this period, the concept of combination of two or more materials to get a more useful material came to light. Then, the researchers started finding lighter but stronger and stiffer materials. Actual research started for composite materials at the beginning of 1960's. Almost 75% of all research and development work in composites has been done since 1965. Then, there has been an ever increasing demand for materials even stiffer and stronger but lighter in all fields of engineering and non-engineering. The overall performances of composite materials are so great and diverse that no one material is able to satisfy the combined properties. This naturally directed to a reappearance of the ancient concept of combining different materials to form a composite material to satisfy the user requirements. Such composite material systems result in a performance better than its constituents and they offer the great advantage of a flexible design. Thus, in principle, one can tailor make material as per specifications of an optimum design of composite structrues.

The composite materials have introduced flexibility to design engineering. In addition, forcing the designers and analysists, to create a different material for each application, this leads to savings in weight and cost of the structures. The main reason for more increasing use of composite materials is that, our society has become more energy conscious. This factor led to an increasing demand for light weight in addition to stronger and stiffer structures in all fields and composite materials and increasingly providing answers shown in Fig.13.1. The Fig. 13.1 shows the comparison of steel, aluminium and composite materials for different mechanical properties. In the Fig. 13.1 given below, the properties for different materials such as Steel, Aluminium and Composites, which are represented in the bar chart as Series 1, Series 2 and Series 3 respectively.

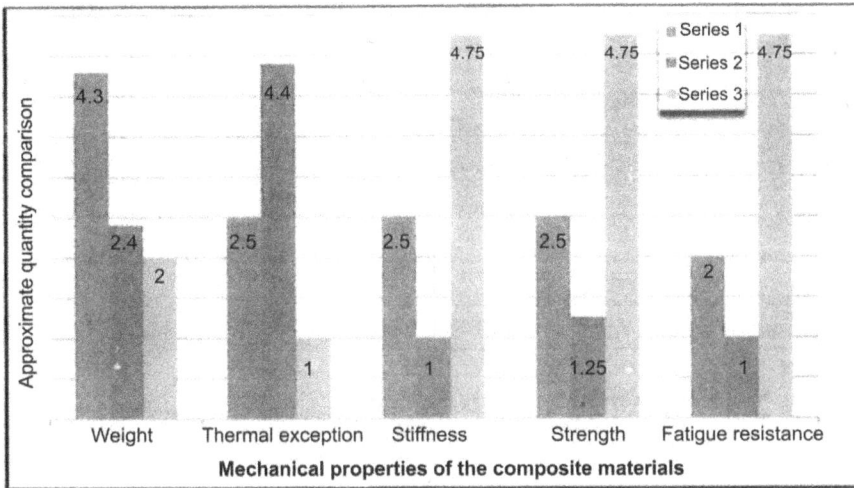

Fig. 13.1 Comparison of the mechanical properties of conventional materials and composite materials

Fig. 13.1 represents the possibility of improvement in several mechanical properties by the use of composite materials compared to conventional materials.

The composites construction made of glass fibre reinforced resins have been in use since 1920. These are very light and strong, even though their stiffness (modulus) is not very high, this is mainly due to the glass fiber itself is not very stiff. The emergence of the advanced fibers of extremely high modulus, for example, boron, carbon, silicon carbide and alumina took place during the year 1970. These fibers have been used for reinforcement of resin, metal and ceramic matrix materials. The fiber reinforced composites are stronger and stiffer in the fibrous form than in any other form such as bulk form, crystal form, thick rod form etc. We can arrange fibers in any direction say one-direction, two-directions, or three-directions, still fibers are stronger and stiffer in fiber direction than in any other directions.

Specific strength and specific modulus of some wood materials are given in Table 13.1.

Table 13.1 Mechanical properties of few wood materials

Wood	Density (r) (Kg/m³)	Young's Modulus (E) (Mpa)	Tensile Strength (s) (Mpa)	Specific Modulus (E / p) (Mm)	Specific Strength (s/r) (Mm)
White Oak	680	12,300	5.4	1.8088	7.941
Paper Birch	550	11,000	–	2.00	–
Douglas Fiber	500	12,500	2.4	2.5	4.8
White pine	380	10,100	–	2.65	–
Red Wood	400	9,200	1.7	2.3	4.25

WHAT IS COMPOSITE MATERIAL?

It is a fact that, practically everything in this world is a composite material ranging from living things to non-living things. Hence, a common piece of metal is a composite (polycrystal) and many grains (or single crystals).

Definition of Composite Materials

The material which satisfies the following conditions are called composite materials

(a) It should be manufactured and tailored according to the requirements.

(b) It is a combination of two or more materials which are physically and/or chemically distinct, suitably arranged or distributed phases with an interface, separating them.

(c) Characteristics of composites are different than its constituents such as fibers and matrices.

In a broader sense composite materials means two or more materials are combined on a macroscopic scale to form a useful material. The main advantage of composite materials is that they usually exhibit the best qualities of their constituents and often better qualities that neither constituent possesses.

WHY DO WE NEED COMPOSITE MATERIALS?

Composite materials provide phenomenally high quality materials compared to conventional materials. Based on applications we can tailor make composites to improve the properties required for specific applications. Composites are good at providing better properties, the properties that can be improved by construction of composite materials compared to conventional materials are

- High strength
- High stiffness
- Very good fatigue life
- Less weight
- Corrosion resistance
- Wear resistance
- Temperature dependent behavior
- Attractiveness
- Acoustical insulation
- Thermal insulation, etc.

These are some properties, which are improved by making composite materials. All these properties cannot be improved with one composite construction and there will be no such requirement in the industry. Depending on the industry requirements the corresponding properties can be improved.

HOW COMPOSITES ARE USEFUL COMPARED TO CONVENTIONAL MATERIALS?

As mentioned earlier, technological development depends on the advances in the field of materials. In this direction, the invention of composite materials is a noble achievement. Composite materials help in optimum utilisation of materials for the given application and loading conditions. Use of composite materials for many applications helps in reducing the wastages up to 50%. As an example, consider a tapered beam has to be made from conventional material and composite material, here, it requires to machining out the extra material if conventional material is used, whereas, there is no need to remove the material if composite material is used instead the ply drop technique will be adopted for tapered beam construction.

In addition, FRC (Fiber Reinforved Composites) are stronger and stiffer in the fibrous form than in any other form such as bulk form, particle form, etc. Material in bulk form will be weak because it may consist of some flaws, voids, foreign particles etc. So that when a material is subjected to external loadings stress concentration will be more at the location of flaws, voids, foreign particles etc. Thus, failure may start initiating at these locations and may propagate further and lead to catastrophic failure of the structures. When we draw fibres from the bulk material the chances of fibres having flaws, voids, foreign particles etc. are very less compared to the bulk materials. So, fibers are stronger and stiffer in longitudinal direction than in transverse direction. Hence, composite materials are more useful than the conventional materials.

CHARACTERISTICS OF COMPOSITE MATERIALS

Composite materials are made up of one or more discontinuous phases embedded in a continuous phase. The discontinuous phase is generally harder and stronger then the continuous phase and is called the reinforcement or reinforcing material, whereas the continuous phase is called the matrix or resin.

It is true that, properties of composites are strongly influenced by the properties of their constituent materials, their distribution, and the interaction among them. The composite properties mainly depend on – may be the volume fractions and sum of the properties of the constituents. In addition the geometry of the reinforcement may be described by the shape, size and size distribution. The reinforcement in composite material may differ in concentration, concentration distribution and orientation, depending on the type of fibers and resin materials used.

For the purpose of convenience, shape of the discrete units like fibers, flakes etc., of the discontinuous phase may often be approximated by Spheres or Cylinders, these are some natural materials such as mica and the clay-minerals. Some man made materials such as glass flakes that can be described as rectangular cross sectioned prisms or platelets etc.

1. Concentration is measured in terms of volume or weight fractions. It is the single most important parameter influencing the composite properties. In addition, it is an easily controllable manufacturing variable used to alter the properties of the composite materials. So that, it is necessary to have optimum concentration to get the high quality composite materials, that is the ratio of fiber and matrix

should be around 65% and 35% respectively. These values differ for different fibers and matrices.

2. Concentration distribution is a measure of homogeneity or uniformity of the composite materials. The homogeneity is an important characteristic that determines the extent to which a representative volume of material may differ in physical and mechanical properties compared to the average properties of the material. The fibers and matrix distribution in a composite material should be optimized. The concentration distribution should be maintained properly to avoid clustering of fibers. Non-uniformity of the composites should be avoided because it reduces those properties that are governed by the weakest link in the material. As an illustration, failure in a non-uniform material will initiate in the region of lowest strength, thus adversely affecting the overall strength of the material. Thus, the fibers and the matrix distribution in a composite materials is a key parameter.

3. Orientation of the fibers in a composite material plays a very important role. Orientation of the reinforcement affects the isotropy of the composite materials. When the reinforcement is in the form of particles, with all their dimensions approximately equal then, this lead to isotropic and some lead to anisotropic. In addition to this, for fiber-reinforced composites, such as unidirectional or cross-ply composites, anisotropy may be desirable. In a fiber reinforced composite materials orientation of the fibers according to the designer requirement has to be maintained. The variation in the fiber orientation allowed is ± 2°. Variations more than the prescribed orientation will lead considerable reduction in strength and stiffness of the composite structures. If the fiber orientation crosses the limit, then the structure may not be able to carry the load which is designed for and may lead to catastrophic failure. Thus, orientations of the fibers have to be maintained properly.

CLASSIFICATION OF COMPOSITE MATERIALS

Composite materials can be classified in many ways, based on fibers, particles, laminates, flakes etc. The composite materials are broadly classified into fiber reinforced and particle reinforced composites mainly based on whether the composites construction is made of fibers or particles. Then, the fiber reinforced composites can be classified into single layer and multilayer composite depending on the composite structure is made up of single or multi-layer construction. In the same way, particle reinforced composites can be classified into random and preferred orientations, depending on whether the particles are placed randomly or in preferred directions in a composite construction.

Further, towards classifications, single layer composites can be classified into continuous fiber reinforced and discontinuous reinforced composites based on whether the fiber used for composite construction are continuous or discontinuous. Then, coming to the classification of multi-layered composite construction, here it is classified into laminated and hybrid composites depending on whether the composite structure is laminated or hybrid construction. In this direction, further continuous fiber reinforced composites can be classified into unidirectional reinforcement and bidirectional reinforcement based on the fibers used in the laminate construction are unidirectional or bidirectional layup. In continuation, discontinuous fiber reinforced composites are classified into random and preferred orientations depending on the fibers placed in the composite are random od preferred orientations. Typical classification of composites is shown in Fig. 13.2.

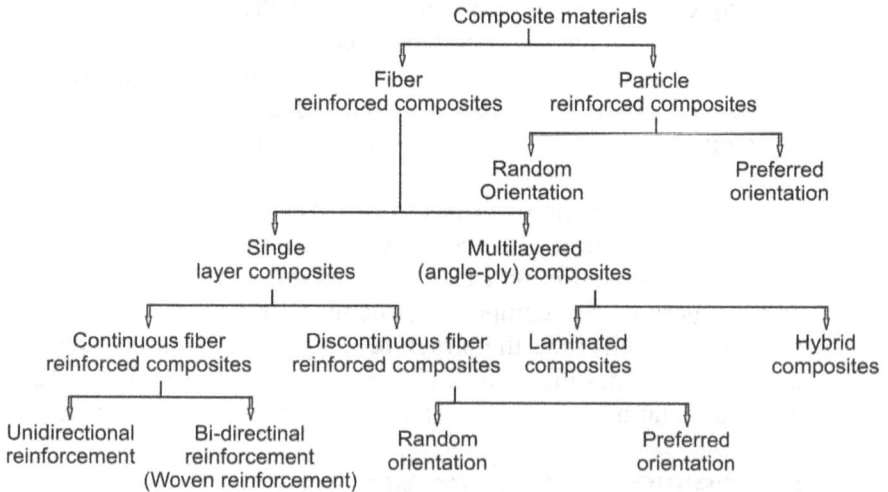

Fig. 13.2 Classifications of composite materials

In addition to the above classifications, composites are broadly classified according to their use. There are five commonly accepted types of composite materials, which are:

1. Particulate composite materials.

2. Flake composite materials.

3. Fiber reinforced composite materials.

4. Laminated composite materials.

5. Combinations of some or all of the first four types.

These types of composite materials are described and discussed in the following sub-sections:

Particulate composite materials

The particulate composite materials consist of particles of one or more materials mixed in a matrix of different material. The particles and matrix are either metallic or non-metallic. The different combinations of these constituents are explained in the following paragraphs.

(a)   **Non-metallic particles in non-metallic matrix composite materials:** The most common example of this type of composite materials is concrete. Concrete has particles of sand and gravel (i.e., rock particles). That are bonded together with a mixture of cement and water (i.e., matrix materials) that has chemically reacted and hardened. The strength of the concrete is generally that of the gravel because the cement matrix is stronger than the gravel.

(b)   **Metallic particles in non-metallic matrix composite materials:** The most common types of these types of composite materials is the metal flakes in a suspension. For example, aluminium paint is actually aluminium flakes suspended in paint. When applied, the flakes orient themselves parallel to the surface and give very good coverage. Similarly, silver flakes can be used to give good electrical conductivity. Cold solder is metal powder suspended in a thermosetting resin. Such a composite is hard and strong and conducts heat and electricity. Inclusion of copper in a matrix increases its conductivity. In a mixture of resin, many metallic additives compared to plastic increase the thermal conductivity, lower the thermal expansion and decrease wear.

(c)   **Metallic particles in metallic matrix composite materials:** A metallic particle in a metallic matrix does not dissolve, as in the case of alloys. Generally lead particles are used in copper alloys and steel to improve the machinability. One more example of type is lead acts a natural lubricant in bearing made from copper alloys.

(d)   Non-metallic particles in metallic matrix composite materials:
Ceramics which are non-metallic particles can be suspended in a
metal matrix; such a composite material is called a cermet. The
cermets can be classified as oxide based and carbide based
composite materials.

The oxide based cermets can be either oxide particles in a metal
matrix or metal particles in an oxide matrix. Generally, this type
of cermets is used in tool making and high temperature
applications where erosion resistance is needed.

The other type, carbide based cermets have particles of carbides
of tungsten, chromium, and titanium. Tungsten carbide in a
cobalt mixture is used in machine parts requiring very high
hardness such as wire-drawing dies, valves, etc.

Flake composite materials

Generally used flake composite materials are glass, mica, metals and
carbon. The size, shape and material of the flake and the amount of
matrix material to be used depend on the type of application. The matrix
materials such as plastics, metals or epoxy resins used in a flake
composite may make up the bulk of the composite or be little just
sufficient to provide bonding of the flakes. For illustration, for decorative
purpose, aluminium flakes are used in molded plastic parts. For this type
of composite, the plastic makes up the bulk of the composite.

Fiber reinforced composite materials

It is natural that, long fibers in various forms are inherently much stiffer
and stronger than the same material in bulk form. For instance, ordinary
plate glass factures at stresses at only 20 MPa, yet glass fibers have
strengths of 2800 MPa to 4800 MPa in commercially available forms and
about 7000 MPa in laboratory forms. Then, it is obvious that, the
geometry and physical make-up of a fiber are somehow very important
for the evaluation of its strength and must be considered in structural
applications. Generally, the inconsistency in a fiber having different
properties from the bulk form is due to the more perfect structure of a
fiber. In case of fibers, the crystals are aligned along the crystal axis. In
addition, there are fewer internal defects in fibers than in bulk material.
As an illustration, in materials that have dislocations, the fiber form has
fewer dislocations than the bulk form.

Properties of fibers: Geometrically, a fiber is characterized not only by
its very high length to diameter ratio but by its near-crystal-sized

diameter. Specific strengths and specific stiffness of a few selected fiber materials are presented in Table 13.2.

where,

$$\text{Strength of the material} = \frac{\text{Strength of the material}}{\text{Density of the material}}$$

$$\text{Specific Stiffness} = \frac{\text{stiffness (modulus) of the material}}{\text{Density of the material}}$$

The more commonly used structural materials, aluminium, titanium and steel are listed for the purpose of comparison. Conversely, a direct comparison between fibers and structural metals are not valid because fibers must have a surrounding matrix to perform in a structural member, but structural metals are 'ready-to-use'. Note that specific strength and specific stiffness are indictors of effectiveness of a fiber, especially in weight-sensitive application such as aircraft and space applications.

Table 13.2 Properties of the fibers or wires*

Fiber or wire	Density (KN/m^3)	Tensile Strength (GN/m^2)	Specific Strength (Mm)	Tensile stiffness (E) $(G\,N/rn^2)$	Specific Stiffness (Mm)
Aluminium	26.3	0.62	24	73	2.8
Titanium	46.1	1.9	41	115	2.5
Steel	76.6	4.1	54	207	2.7
E-Glass	25.0	3.4	136	72	2.9
S-Glass	24.4	4.8	197	86	3.5
Carbon	13.8	1.7	123	190	
Beryllium	18.2	1.7	93	300	16
Boron	25.2	3.4	137	400	16
Graphite	13.8	1.7	123	250	18

* Adopted from Jones

Fibers of graphite and carbon are of great interest in today's composite structures. These fibers are made from rayon, pitch, or PAN (Polyacrylonitrile) precursor fibers that are heated in an inert atmosphere to about 3100°F (1700°C) to carbonize the fibers. For obtaining graphite fibers, the heating exceeds 3100°F for particularly graphitizing the carbon fibers. In addition, as the processing temperature increases the fiber modulus increases, but the strength often decreases. The fibers are far thinner than human hairs, do they can be bent quite easily and hence, carbon and graphite fibers can be woven into fabric. On the other hand, boron fibers are made by vapor deposition technique, by vapor depositing boron on a tungsten wire and coating the boron with a thin layer of boron carbide. Boron fibers are about the diameter of mechanical pencil lead, so that, they cannot be bent or woven into fabric.

A laminate made of layers of at least two different materials that are bonded together is known as laminated composite materials. Lamination and bonding material are used to combine the best aspects of the constituent layers, and to achieve a more useful material. The properties that can be improved by lamination are strength, stiffness, low weight, corrosion resistance, wear resistance, beauty or attractiveness, thermal insulation, acoustical insulation, etc. For illustration, laminates can be made of bi-metals, clad metals, laminated glass, plastic-based laminates and laminated fibrous composite materials, which are described in the following paragraphs.

(a) **Bi-metals:** Laminates of two different metals that usually have significantly different co-efficient of thermal expansion are known as bi-metals. Temperature change make bimetals warp or deflect by a predictable amount, so that, they are well suited for use in temperature-measuring devices. For illustration, a thermostat can be made from a cantilever strip of two different metals bonded as shown in the Fig. 13.3.

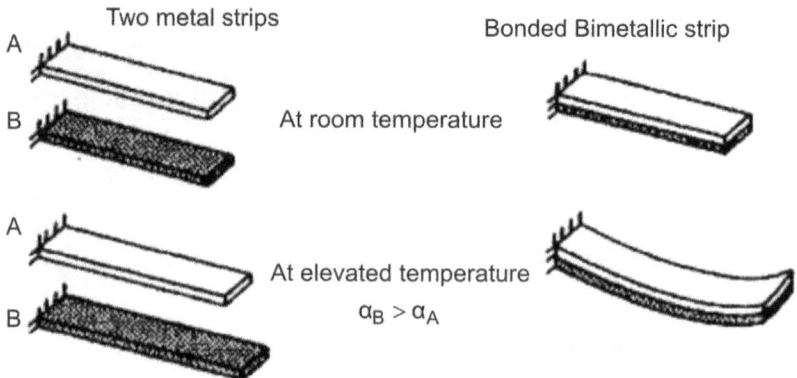

Fig. 13.3 Cantilevered bimetallic strip (thermostat)

Here, metals A and B have respectively thermal co-efficient of expansion α_A and α_B with α_A greater than α_B. Consider two cases

(i)   Two unbounded metal strips of different co-efficient kept side by side.

(ii) The same two strips bonded together.

For case (i), the two strips retain same length at room temperature, but when they are heated, both strips elongate. For case (ii), the bonded strips maintain original lengths at room temperature, whereas, when heated, strip B expands more than strip A, thus making the bimetallic strip to bend.

(b) **Clad metals:** The cladding or sheathing of one metal with other is carried out to obtain the best properties of both. For illustration, high strength aluminium alloys are not good for corrosion resistance; still, pure aluminium and some aluminium alloys are good for corrosion resistance but relatively weak. Thus, a composite material made of high strength aluminium alloy covered with a corrosion resistant aluminium alloy is very good for both high strength and corrosion resistance applications, which are unique and has attractive advantages over the properties of its individual constituents.

(c) **Laminated composites:** For the application of automotive safety glasses, laminated glasses are used. Ordinary window glass, which is generally used, is quite durable enough to retain its transparency under extremes of weather. But, glass is quite brittle and is dangerous because it can break into many sharp edged pieces. In contrast, a plastic called poly-vinyl-butyral is very tough (deforms to high starins without fracture), although it is very flexible and susceptible to stretching. The safety glass is a layer of poly-vinyl-butyral sandwiched between two layers of glass. The glass in the composite material protects the plastic from damage and provides stiffness and the plastic gives toughness to the composite material. Hence, totally, the glass and plastic material protect each other and lead to a composite material with properties vastly improved over its constituents. In reality, the scratchability property of the plastic is totally eliminated because it is the inner layer of the composite laminate.

(d) **Plastic–based laminates:** Many materials can be saturated with various plastics for a plenty of application. For mica is used as layers of heavy Kraft paper impregnated with a phenolic resin-saturated decorative sheet, in addition, that is overlaid with a plastic saturated cellulose material. To bond the layers together, heat and pressure are used. For dissipating heat, an aluminium layer is placed between decorative layer and the Kraft paper layer.

Composite is made of more than two constituents each making an essential, but different contribution to the resulting composite material, which is known as compound composite material.

(e) **Laminated fibrous composite materials:** Various fiber fabrics are used to make laminates, such as glass, Kevlar, nylon, boron, metals, ceramics, etc. These fibrous fabrics can be laminated with various resins to yield important light weight structural components. Laminated fibrous materials are finding many applications, such as strength to weight, stiffness to weight, impact resistance, thermal resistance, aesthetically pleasing appearance, acoustical, environmental, corrosion resistance, etc.

Combination of some or all of the first four types

Variety of multipurpose composite materials exhibit more than one characteristics of the many classes, fibrous, laminated, particulate, or flake composite materials as discussed above.

For illustration, reinforced concrete is both particulate (i.e., concrete is made of gravel in a cement-paste binder) and fibrous (i.e., steel reinforcement). In addition to this, laminated fiber-reinforced composite materials are truly both laminated and fibrous composite materials.

The laminated fiber-reinforced composite materials are a kind of hybrid class of composites, which has both fibrous composites and lamination techniques. The layers of fiber-reinforced material are bonded together with fiber directions of each layer oriented in different directions, which provides different strengths and stiffnesses of laminate in several directions. Hence, the strengths and stiffnesses of laminated fiber-reinforced composites can be tailored to the specific design requirements of the structural element to be built. Applications of this type of materials are rocket motor cases, boat hulls, building panels, aircraft wing panels and body sections, tennis rackets, golf shaft, etc.

Why composite materials are better compared to conventional Materials

In composites, majority of material is stronger and stiffer in the fiber form than in any other form such as bulk form, particle form, crystal form etc., so that, there is a great attraction of fiber reinforcement. The material in the bulk form will have flaws, voids, foreign body etc., when such a body is subjected to external loads stress concentration will be more in the region of flaws, voids, foreign body etc., so that, failure initiation will take place from such region first then propagation of failure will take place and finally there is a chance of catastrophic failure of the structure. In contrast, fibers are free from such defects (e.g., flaws, voids, foreign body etc.,) because when you draw a fiber from the bulk material chances of having defects in the fibers are very remote and will provide full strength and stiffness to the structure. Hence, fiber reinforced composite materials are better, stronger and stiffer than the same material in the bulk form.

The use of fibers for engineering composite materials is based on three important characteristics, which are as follows:

1. Smaller the diameter better will be the properties, a small diameter with respect to its grain size or other microstructural unit. This lead to attaining higher strength than that of the material in bulk form, this is called the size effect. Smaller the size lesser the possibility of having defects in the material. Fig. 13.3 shows that the strength of the carbon fiber decreases as the diameter of the fiber increases.

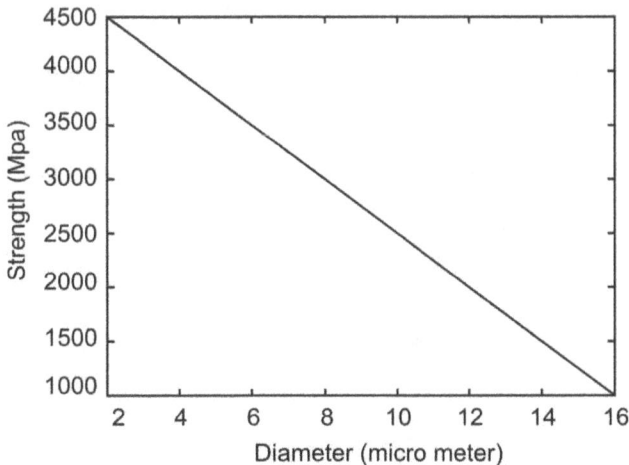

Fig. 13.4 Variation in the strength of the fiber as diameter varies

2. A high aspect ratio, that is length of the fiber to the diameter of the fiber (L/D), which help in applied load to be transferred to the fibers via the matrix material. Thus, long fibers are very good load carrying members and provide very good strength and stiffness to the structure.

3. A very high degree of flexibility that is fibers of very small diameter and very long length are highly flexible, which is the characteristic of a material with high modulus and small diameter.

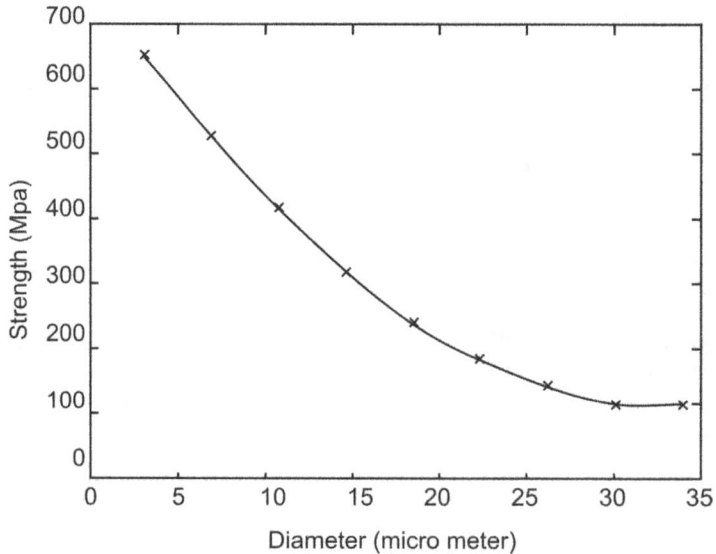

Fig. 13.5 Diameter of fibers of different materials with flexibility equal to that of a 25 µm nylon fiber.

In terms of equation we can represent the measure of flexibility as,

$$\frac{M}{I} = \frac{E}{R}$$

$$MR = EI$$

$$MR = \frac{E \pi d^4}{64}$$

$$\frac{1}{MR} = \frac{64}{E \pi d^4} \qquad .....(13.1)$$

where, M is the bending moment, R is the radius of curvature, E is the elastic modulus, I is the moment of inertia, d is the equivalent diameter.

The inverse of the product of bending moment (M) and radius of curvature (R) as a measure of flexibility. The equa. 13.1 shows that $\left(\frac{1}{MR}\right)$, a measure of flexibility, is a very sensitive function of diameter 'd'.

APPLICATIONS OF COMPOSITE MATERIALS

Composite materials are finding many applications ranging from Aerospace Structures to Aesthetics. Composite materials are finding applications in all fields because of their properties and easy of moulding, flexibility and fabrication technology as shown in Fig. 13.6, with approximate percentages used in different fields. From Fig. 13.6, it is found that the composite materials are finding vast applications in Engineering and non-engineering fields. The composites are used more approximately 31% in Automotive industries, here mainly used for doors, body parts, bumpers etc., different fibers and matrices are used according to their requirements at different parts of the automobile construction. In construction industry that is buildings, bridges, and roads, composites used is approximately 26%, mainly used for construction of beams, columns, slabs, doors, windows, culverts, pavement, etc. Mainly in Japan composites have been used for construction purpose, because Japan is frequent earth quake prone zone, hence it is used for light weight construction and also for ease of dismantling and fabrication.

Marine application is the other major area where approximately 12% of composites have been used. Main components of composites are boat hulls, body etc., this provides light weight structure which helps in carrying more payload and cargo. Approximately 10% of composite materials are used for fabrication of Electronic components. It provides light weight, electrical resistance, aesthetics etc. For Consumer products approximately 8% of composite materials are used, it includes many parts of the all consumer products. In continuation, about 8% of the composite materials are used for fabrication of Appliances. Then, in Aerospace applications about 1% of composite materials are used. Aerospace industry requires mainly light weight with very high strength and stiffness fibers and matrices. Usage of composites is looks very less (about 1%) compared to other Industries, this is because numbers of Aerospace Vehicles are very less compared to Automobile Vehicles. But Aerospace Vehicles are built with almost full composite materials components compared to any other field as mentioned above. Finally about 4% of composites are used for many miscellaneous applications.

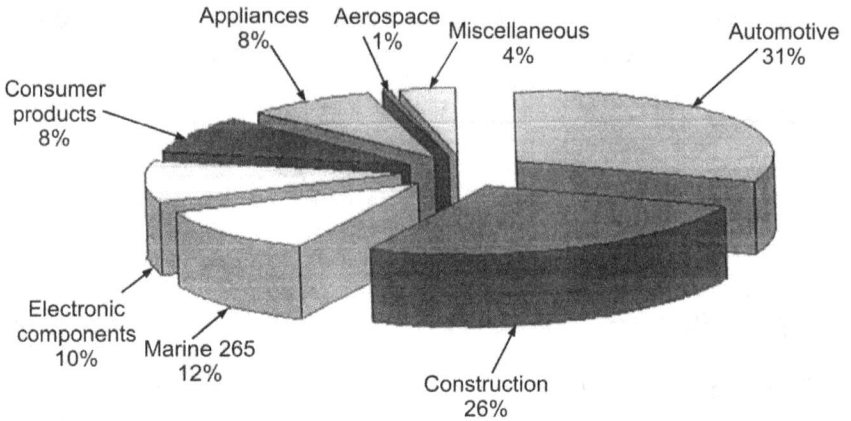

Fig. 13.6 Application of composite materials with approximate percentages in different fields of Engineering and non-Engineering fields.

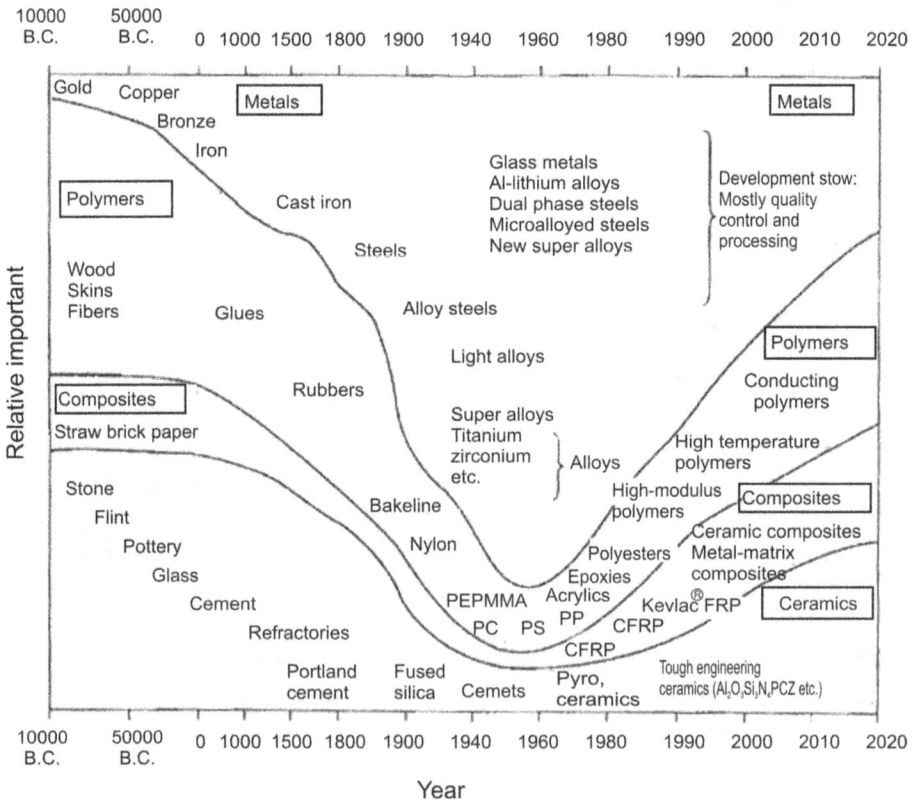

Fig. 13.7 Relative importance of composites and other materials with respect to year

Fig 13.6 shows the relative importance of different materials over the year. Composites technology was used for making bricks during 10000 B.C., then its usage started declining and again taken flight from 1960 onwards and now it is used mainly as fiber reinforced composites and polymers for many advanced composites applications. Hence, from the Fig. 13.6 it is clear that composites use has been increasing over the year and other materials usage has been declining.

COST OF THE COMPOSITES COMPARED TO OTHER MATERIALS

Manufacturing of composite components are very costly because of the very high cost of fabrics made of fibers and resins, The very high cost is mainly due to non-availability of fabrics in India. In India only fabrics made of glass fibers are available, so these are very cheap but again the quality is an important issue. Most of the fabrics made of carbon, boron, Kevlar, ceramic, metal etc., fibers are procured from abroad, so that, the cost of the composite materials are very costly compared to any other materials.

The other main factor for high cost is due to exchange rate of rupee v/s dollar is very high. So, any material manufactured locally will be available cheaply.

The Fig 13.7 shows the comparison of cost of many materials with respect to Young's Modulus of the materials. From the figure it is found that, one is the reasonable cost of the material as shown in the x-axis and any value above one is considered as expensive and below one is considered as cheap. So that, region marked in the plot shows that composites falls in the region of expensive materials compared to other many materials as shown in Fig. 13.8. Similarly, in the y-axis 10 GPa is considered as the moderate value of Young's modulus, above this value is considered as the stiff material and below this value is taken as the flexible materials. Again composite materials fall in region of stiff materials compared to many materials as shown in Fig. 13.8. In the same way we can compare cost with respect to strength, weight, density etc.

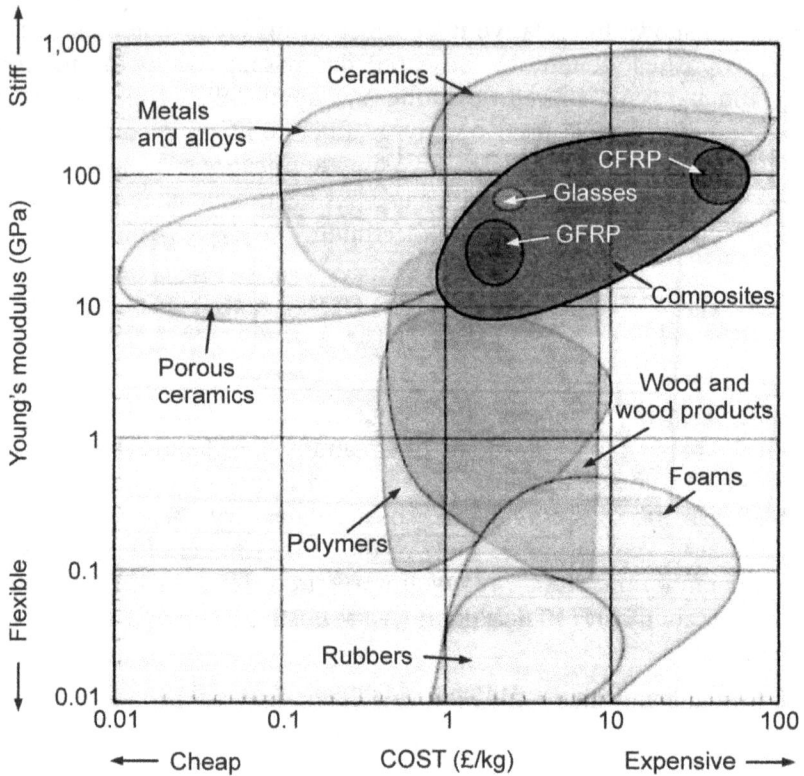

Fig. 13.8 Comparison of cost with respect the Young's Modulus of many materials

STRENGTH AND STIFFNESS PROPERTIES OF THE COMPOSITES

The specific strength and specific stiffness play very important role in composites, because composites are mainly used for construction of light weight structures with very high strength, high stiffness and other improved properties compared to conventional materials. In addition to the property data provided in the Table 13.3 for different materials, Fig. 13.9 gives information about the specific strength and specific stiffness of different materials in different forms such as bulk form, fiber form, unidirectional and bidirectional composites. From Fig. 13.9, it is clear that specific strength and specific stiffness of the metals such as steel, titanium and aluminium all will have almost the same values though their strengths, stiffnesses and densities are different.

Again from Fig.13.9, one can note that, specific strength and stiffness of fibers are very high, because fiber densities are very less and strength and stiffness are very high. However, in applications fibers in the form of woven fabrics only will not be used, these have to be used with resins (matrix material). Resin is a bonding material used for bonding fabrics. Single layer of fabric mixed with resin is known as lamina and two or more layers of fabrics mixed with resin are known as laminates. From the Fig. 13.9, it is found that the specific strength and stiffness of unidirectional lamina is higher than that of biaxially isotropic laminates for all types of fibers. Unidirectional means all fibers in a lamina are oriented in one direction and biaxially isotropic laminate means laminae are oriented in two directions. Hence, it is a fact that specific strength and stiffness are used in composites constructions rather than consideration of only individual properties.

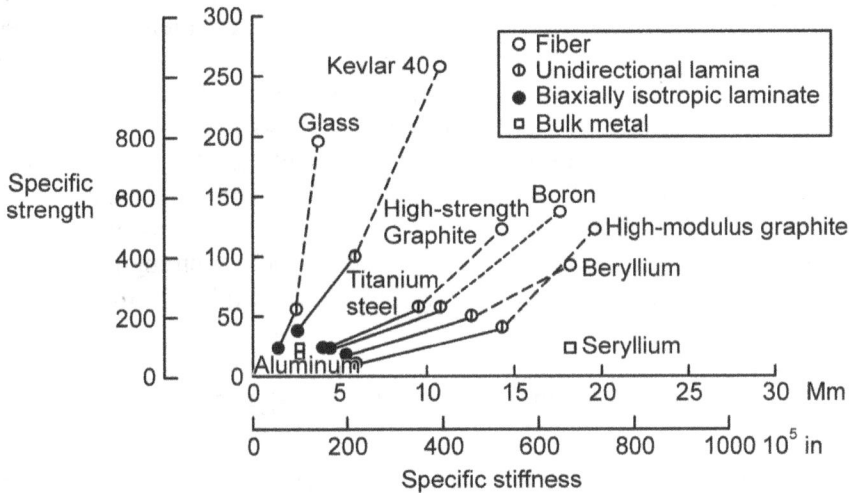

Fig. 13.9 Specific strength and specific stiffness of composite materials and metals

So far we have studied about the details of composite materials, in the following chapter we shall discuss about different types of fibers and matrices and their properties, applications, salient features, merits and demerits.

FIBERS AND MATRICES

In the previous chapter we have studied about the definition of composite materials, its classifications, characteristics, compositions, applications, properties as a whole, cost comparison etc. This chapter contains about the details of different type of fibers and matrices used for the construction of composite materials, their properties, applications, characteristics, classification and many other salient features. Composite are made up of combination of fibers and matrices, different fibers and different matrices have different properties and applications.

Designer is free to select the type of fibers and matrices required for a particular application. There are plenty of possible selections of fibers and matrices are available for a particular application. Thus, the engineer or designer should have the thorough knowledge about different types of fibers and matrices available world-wide and their properties, classifications, characteristics, applications. In addition to these things designer should have complete understanding of merits and demerits of all fibers and matrices, their applicability, drawbacks, limitations etc. Once designer has the complete knowledge about the fibers and matrices, then only a designer can select proper fibers and matrices for a particular application. The fibers and matrices have different density, ultimate stress, Young's modulus, ultimate strain, specific strength, specific stiffness and host of other properties, which are explained in details as follows.

REINFORCEMENTS (FIBERS)

Reinforcements need not to be in the form of long fibers only, they can be in the form of particles, flakes, whiskers, discontinuous fibers, continuous fibers, sheets, etc. It is true that the great majority of materials are stronger and stiffer in the fibrous form than in any other form, so that, there is great attraction of fibrous reinforcements. In this class of fibers, we are interested in advanced fibers which possess very high strength and stiffness with a very low density. Many naturally occurring fibers can be used in situations where high strength and stiffness are not important but the advantage is low cost. The vegetable kingdom is the largest source of fibrous materials, Cellulose fibers in the form of cotton, flax, Jute, hemp, sisal, and ramie have been used in the textile industry, whereas wood and straw have been used in the paper industry as well as for manufacturing of bricks and for construction of bridges, roads, buildings etc. Other natural fibers are hair, wool & silk, consist of different forms of protein. Several fibers are available which are, glass, Kevlar (an aramid), boron, silicon carbide, carbon and alumina, etc. Properties of some of the fibers are given in Table 13.3. These fibers exhibit very high strength and stiffness properties. The following paragraphs describes about the above mentioned fibers.

Table 13.3 Properties of some fibers

Fibers	Density (Kg/rn³)	Diameter (μm)	Tensile Strength (GPa)	Young's Modulus (GPa)	Co-efficient of Thermal Expansion ($\times 10^{-6}$)
Boron	2340	100 – 406	3 – 4	380 – 400	8.3
Carbon	1600 – 2268	5 – 10	3.5	400	8.4
Organic Fibers 1. Kevlar	1400	12	2.8	65 – 125	8
Ceramic Fibers 1. Alumina	3950	15 – 25	1.38	379	
2. Silicon Carbide	3300	140	3.50	430	

MATRICES (RESINS)

Matrices (resins) are bonding materials, without which only reinforcements (fibers) are of no use. Matrices are bonding materials as well as a real load bearing and load transferring materials. The load applied on any composite structure is directly applied on the matrix material and the load will be transferred to the fibers through matrix materials. Thus, proper bonding should be ensured. The strength of the matrix material is equal to the strength of the bonding.

Numerous matrix materials are available for the composite manufacturing such as epoxy matrix material, ceramic matrix materials, metal matrix materials, cement matrix materials, carbon matrix materials etc. The Designer has to select the proper matrix materials based on its properties, functions, feasibility etc., for a particular application. The selection of matrix materials is mainly based on its merits and demerits and the application for which it is to be used. So that, the designer should have proper knowledge about the different matrices available world-wide and their properties, application for which a particular matrix material to be used, functions, feasibility etc. The following paragraphs describe about these matrices.

TYPES OF FIBERS (REINFORCEMENTS)

There are different type of fibers are available world-wide; here few fibers are explained in brief about their properties, applications, merits, demerits and salient features of each.

Glass fibers

Glass fiber, which is a generic name similar to carbon fiber or steel. Generally, glass fibers are silica based (~ 50-60% SiO_2) and contain a host of other oxides of calcium, boron, sodium, aluminium and iron. The composition of some commonly used glass fibers are presented in Table 13.4. The nomenclature E stands for electrical, because E glass is a good electrical insulator, in addition to having good strength and Young's modulus, C stands for better resistance to chemical resistance, S stands for higher silica content and is able to withstand high temperatures than other glasses.

Table 13.4 Chemical composition of some glass fibers (wt. %)

Composition	E-Glass	C-Glass	S-Glass
SiO_2	55.2	65.0	65.0
Al_2O_3	8.0	4.0	25.0
CaO	18.7	14.0	–
MgO	4.6	3.0	10.0
Na_2O	0.3	8.5	0.3
K_2O	0.2	–	–
B_2O_3	7.3	5.0	–

Properties and applications

Typical properties of E-Glass are:

Density–2550 Kg/m³; Tensile strength–1750 MPa; Young's Modulus – 70 GPa; Co-efficient of thermal expansion $\sim 4.7 \times 10^{-6}$ (K^{-1}).

The density is quite high and the specific modulus is moderate. Glass fibers are used for reinforcement of polymer, epoxy and phenolic resins. Glass fibers are cheap and available in a variety of forms such as continuous strand, roving, chopped fibers, fabric, etc. Continuous strand is a group of 204 individual fibers; roving is a group of parallel strands; chopped fibers consist of strand or roving chopped to lengths between 5 and 50 mm. Also glass fibers are available in the form of woven fabrics or nonwoven mats.

Moisture reduces the strength of the glass fibers. Glass fibers are susceptible to static fatigue, that is, they cannot withstand load for longer duration.

Glass fiber reinforced composites are used extensively in the building and construction industry. Generally, these are called glass reinforced plastics (GRP). GRP are utilized in the form of a cladding for other structural materials or as an integral part of a structural or non-load bearing wall panel. Many other applications are window frames, tanks, bathrooms units, pipes, ducts, boat hulls, chemical industry (e.g., as storage tanks, pipelines, and process vessels), rail & road trains, port industry, Aerospace Industry etc.

Boron fibers

Boron is by nature a brittle material. It is made by chemical vapor deposition of boron on a substrate. The boron fiber produced by this method is itself a composite fiber. It is a fact that high temperatures are required for this deposition process. The choice of the substrate to obtain boron fiber is limited, generally, fine tungsten wire, or boron can be used for this purpose. The some properties of boron fibers are given in Table 13.1.

Properties and applications

Boron fibers, by nature are compositses. They exhibit complex internal stresses and defects such as voids and structural discontinuities result from the presence of a core and the deposition process. So, that, the strength of boron fiber would be different than the strength of boron. The tensile strength of boron fiber is 3-4 GPa and the Young's modulus is 380-400 GPa. The density of the Boron fibers is 2340 Kg/m^3, which is about 15% less than that of aluminium. It's melting point is 2040°C and thermal expansion coefficient of 8.3×10-6°C' up to 315°C. The flexural strength of the boron is 14 GPa.

Application of boron fiber composites are not much because of very high cost. Still, these are used in U.S. Military Aircraft, such as F-14 & F-15 and in the U.S. Space Shuttle. Generally, boron fibers are used for stiffening golf shafts, tennis rackets, and bicycle frames, etc.

Carbon fibers

Carbon is a very light element and its density is 2268 Kg/m^3. It can exist in a variety of crystalline forms. Two common forms of carbon are graphite structure and covalent diamond structure. The carbon atoms arranged in the form of hexagonal layers are called graphite structure, and those in a three dimensional configuration with little structural flexibility are known as covalent diamond structure. Carbon in the graphite form is highly anisotropic with longitudinal young's modulus of about 1000 GPa and transverse Young's Modulus of about 35 GPa. The properties of carbon fibers are shown in Table 13.2.

Properties and applications

The density of the carbon fiber varies with the precursor and the thermal treatment and also other properties also varies as shown in Fig. 13.10. Based on this carbon fibers are classified as follows.

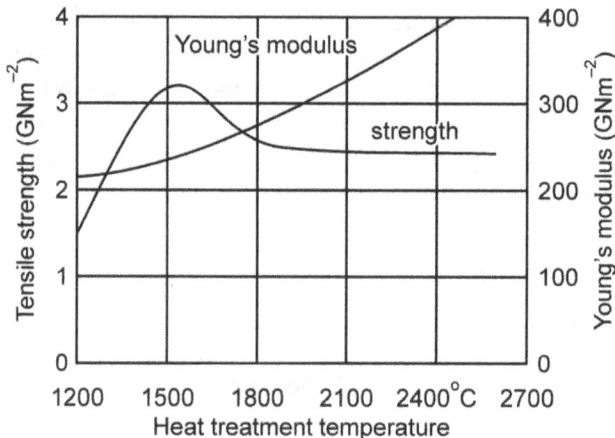

Fig. 13.10 Variations in tensile strength and Young's modulus of carbon fibers as the heat treatment temperature varies.

Based on this carbon fibers are classified as follows.

Classification of carbon fibers

1. High tensile strength but medium young's modulus (HT) fibers (2–3 GPa)
2. High Young's modulus (HM) Fibers (400 GPa)
3. Extra or Super high tensile strength (SHT) fibers (>5 GPa)
4. Super high modulus type (SHM) fibers (>400 GPa)

Applications

1. Good Electrical Conductors
2. Aerospace, sporting goods industry,
3. Cargo bay doors and booster rocket casting in the U.S. Space Shuttle.
4. Turbine, Compressor, Windmill blades flywheels
5. Medical equipment, & implant materials (e.g., Ligament replacement in knees & hip joint replacement).

Organic Fibers

In organic fibers strong covalent carbon-carbon bond exists. The fact that the covalent carbon-carbon bond is a very strong one, so linear chain polymers such as polyethylene to be potentially very strong & stiff. Organic fibers have different names based on chemical composition, type of polyamide, and synthetics used for making particular fibers. Aramid fiber is a general name for a class of synthetic organic fibers called aromatic polyamide fibers and these are generally called Kevlar. Nylon is a generic name for any long-chain polyamide. Kevlars are rigid rod like polymers, strong covalent bond in the fiber direction and weak hydrogen bonding, in the transverse direction; these exhibit anisotropic properties.

Kevlar fiber has poor characteristics in compression; its compressive strength is 1/8 of its tensile strength. In tensile loading conditions, the load is carried by the strong covalent bonds, while in compression loading conditions, due to weak bonding in compression (like chain, weak hydrogen bonding & vanderwalls bonds come into play) which lead to easy yielding, buckling & kinking of the fiber. In situations of compressive loading, it is better to use hybrid composites, wherein combination of Kevlar with carbon or any other fibers in matrix is used. Carbon takes the compressive loads and Kevlar takes the tensile loads.

Kevlar fiber being an organic fiber undergoes photo degradation when exposed to light (both visible as well as ultraviolet), which leads to discoloration loss in mechanical properties. This can be minimized by adequately coating the Kevlar composite surface with a light absorbing material.

Classifications and applications

Kevlars are classified based on their compositions used foe the manufacturing and particular type of kevlar fibers are used for particular applications. These are classified as follows.

(i) **Kevlar:** This is used for rubber reinforcement for tires, belts, rubber goods etc.

(ii) **Kevlar 29:** This is mainly used for ropes, cables, coated fabrics, architectural fabrics, ballisore protection fabrics etc.

(iii) **Kevlar 49:** This is suitable for Aerospace, Marine, automotive, sports industries etc.

The properties of Kevlar and Kevlar 29 are same whereas Kevlar having better durability. The properties of Kevlar 29 and Kevlar 49 and given in Table 13.5.

Table 13.5 Mechanical properties of kevlar29 and kevlar 49

Properties	Kevlar 29	Kevlar 49
Density (Kg/m^3)	1440	1440
Diameter (μm)	12	12
Tensile Strength (GPa)	2.80	2.80
Strain to Fracture (%)	4.00	2.3
Young's Modulus in Tension (GPa)	65	125

Ceramic Fibers

Continuous ceramic fibers pocesse high strength and elastic modulus with high-temperature capability and are free from environmental attack. These are very attractive for high temperature structural material applications.

There are several methods for the fabrication of fibers. Ceramic fibers are fabricated by three different methods as follows.

(i) Chemical vapor deposit

(ii) Polymer pyrolysis

(iii) Sol-gel techniques.

Ceramic fibers are formed by using different combinations of materials and different names are coined for each fibers based on the maximum composition of a particular material in the fibers. For example, alumina fibers means the fibers made of above 50% of the contents in the fibers consists of alumina and remaining contents will be made of other oxides such as carbon, oxygen, calcium, boron, sodium, aluminium, iron etc. Likewise, different names are coined for different fibers based on maximum content of a particular oxide in the fibers. Different types of ceramic fibers are as follows.

(a) **Alumina fibers:** The strength of these fibers will remain the same at high temperatures. The properties of the fibers are shown in Table 13.6.

(b) **Alumina & Silica Fibers (mixture):** These are very stiff fibers and the strength of these fibers will remain the same at high temperatures. The properties of these fibers are almost the same as properties mentioned in Table 13.6.

Table 13.6 Properties of the alumina fibers

Diameter (μm)	Density (Kg/m^3)	Tensile Strength (MPa)	Young's Modulus (GPa)	Melting point (/$^\circ$C)
15-25	3950	1380	379	2045

(c) **Silicon Carbide fibers:** These are stiff, strong and available cheaply and abundantly. These are available in the form of whiskers and particles. Normal available diameter of whiskers will be 5-10 μm and the length will be 10-50 mm and the l/d ratio will be from 50 to 10,000. Whiskers do not have uniform dimensions or properties. Whiskers have very high Tensile strength and high Young's Modulus which are typically 8.4 GPa and 581 GPa respectively.

In spite of having very good properties, there are many disadvantages of whiskers

(i) Properties of whiskers are not constant and they vary largely.

(ii) Handling of whiskers and alignment in a matrix to produce composite is very difficult.

(iii) Whiskers have non uniform dimensions etc.

(d) **Silicon nitride:** These fibers have very good properties like mentioned for silicon carbide fibers. The diameter of these fibers vary largely say in the range of 5-50 μm. These are very expensive materials.

(e) **Boron Carbide:** These are very light say density ranges between 1000 kg/m³ 1200 Kg/m³. These fibers are very strong and stiff say tensile strength and Young's modulus are in the range of 6 GPa and 480 GPa respectively

(f) **Boron nitride:** These are like carbon fibers, they are very strong and stiff. The density of these fibers will be around 2200 kg/m³ and the tensile strength and stiffness are in the range of 5 GPa and 400 GPa respectively. These fibers provide greater oxidation resistance. They exhibit excellent dielectric properties and other host of many good properties.

Metallic Fibers

Generally many metals in the form of wires show high strength levels. Obtaining of small diameter wires is very difficult. Metals in the bulk form may contain flaws, voids, foreign body etc., due to these things. When the body is subjected to external loads the stress concentration could occur at these regions and may lead to catastrophic failure. But the metallic fibers are free from these defects and exhibit more strength and stiffness compared to its bulk form. The great advantage of metallic wires is that they show very consistent strength values than any of the ceramic fibers. Fibers can be obtained from all metals, most important wires are:

1. **Beryllium:** This has low density with high modulus. The typical properties of these fibers are Young's modulus is 300 GPa, density is 1850 Kg/m³ and strength is 1.30 GPa. These fibers are toxic and are very costly. Properties of this metal are given in Table 13.7.

2. **Steel:** It provides very high strength and low cost. It is a common reinforcement material for concrete and tyres. Very fine (0.1 mm diameter) & high carbon (0.9%) steel wires have very high strength values (~ 5Gpa), which are low in toughness. These fibers show consistent values of Young's Modulus, 210 GPa and density, 7850 kg/m³. Production of fine metallic wires is very expensive. The cost increases if diameter is less than 100 mm. Properties of this metal are given in Table 13.7.

3. **Tungsten:** This shows very high modulus and refractory (unmanageable) properties. These wires were originally developed for lamps. It has a very high density, 19300 Kg/m^3 but because of its refractory properties, it has been used in some nickel and cobalt based Super alloys. It is used in electrical industries. The disadvantage is that, its ease of oxidation and the oxide of tungsten is likely to volatilize at high service temperatures. Properties of this metal are given in Table 13.7.

Table 13.7 Typical properties of some metallic wires

Material	Diameter (μm)	Density (Kg/m^3)	Tensile Strength (GPa)	Young's Modulus (GPa)	Co-efficient of thermal expansion (10^{-6} K^{-1})	Melting Point (/ °C
Steel (0.9% C)	100	7850	4.25	210	11.8	1300
Stainless Steel (18-8)	50-250	8000	1.00	198	18.0	1300
Beryllium	50-200	1850	1.26	300	11.6	1280
Tungsten	25-250	19300	3.85	360	4.5	3400

TYPES OF MATRIX (RESIN) MATERIALS

There are several types of matrix materials are available world-wide; here few matrix materials are explained in brief about their properties, applications, merits, demerits and salient features of each. Different matrix materials are as follows.

Polymers

These are structurally more complex than metals or ceramics. They are cheap and easily processible. Conversely, polymers have lower strength and modulus and lower temperature resistant limits. Prolonged exposure to ultraviolet light and some types of solvents can lead to the degradation of properties. These are generally poor conductors of heat & electricity, because of the predominant covalent bonding. However, these are more resistant to chemicals than metals. The polymers are structurally giant chain like molecules (thus, the name macromolecules is also used) with covalent bonded carbon atoms. The process of joining many monomers (molecules) together to form polymers is called polymerization. These polymers exist in different chain configurations such as linear, branched, cross linked and ladder, which are explained briefly as follows,

Different Chain Configuration of Polymers

(i) **Linear Polymers:** Polymers consist of a long chain of atoms with attached side groups as shown in Fig. 13.11. Examples of this category are Polyethylene, polyvinyl chloride, and poly methyl methacrylate. In this type coiling and bending of the chain could occur.

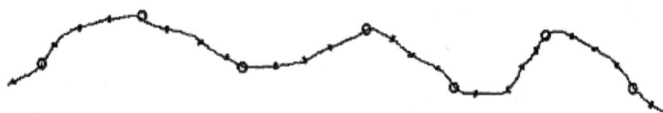

Fig. 13.11 Linear chain polymers

(ii) **Branched Polymers:** Polymer branching can occur with linear, cross linked or any other types of polymers as shown in Fig. 13.12.

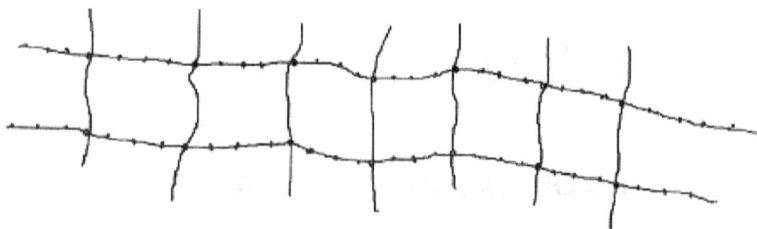

Fig. 13.12 Branched chain polymer

(iii) **Cross linked polymers:** The molecules of one chain are bonded with another (see Fig. 13.13) cross linking results in a 3-D network. Cross linking helps in avoiding sliding of molecules one another difficult, and makes the polymers strong & rigid.

Fig. 13.13 Cross linked chain polymers

(iv) **Ladder Polymers:** Two Linear polymers linked in a regular manner are called ladder polymers as shown in Fig. 13.14. These are more rigid than linear polymers.

Fig. 13.14 Ladder chain polymer

Polymers can again be classified based on their properties and mechanical behaviour as thermoplastics and thermosets.

(i) **Thermoplastics:** Most linear polymers soften or melt on heating, these are called thermoplastic polymers. These are suitable for liquid flow forming. Examples of thermoplastic polymers are Polyethylene, Polystyrene and Polymethyl methacrylate.

(ii) **Thermosets:** Cross liked polymers do not soften on heating, these are called thermosetting polymers. These polymers decompose on heating. Examples of thermosetting polymers are:

1. Rubber cross linked with sulphur, that is, vulcanized rubber. It has 10 times the strength of the natural rubber.

2. Phenolic, polyester, polyurethane and silicone.

CHAPTER 14

Hardness

Hardness is a property exhibited by all materials. If we hit a small block of copper with a hammer, the block elongates and a dent of the shape of the hammer tip can be seen on the copper block. From this we can say that the hammer is harder than the copper block. Similarly we can take any two materials and find out which of them is harder. This can be done by hitting one with the other, by rubbing one against another, or trying to pierce or dent one with the other. In all the cases, the harder material remains unaffected. Depending upon the way in which the hardness is found out, the method is suitably termed. When we try to break one of the two materials by sudden loading or impact, it is called impact hardness. The word impact hardness is modified in technical usage and termed as rebound hardness. Particularly for metals which are crystalline, a sudden fracture is not possible but the metal surface gives a rebound to the material which strikes it. The amount of rebound is proportional to the hardness of the surface. In fact, the rebound hardness is measured thus.

Scratch or wear hardness is that property of materials by virtue of which they resist wear or abrasion. Similarly, the amount of resistance offered by a metal for indentation or penetration is defined as the indentation hardness of the metal. Thus, we define hardness as the property of a metal by virtue of which it resists scratch, wear, abrasion or indentation.

In the industry, the indentation hardness is most commonly measured. In the metallurgical sense, it has become a practice to understand hardness as the indentation hardness only, unless otherwise specified. Rebound hardness, too, is often employed for routine check-up. Unfortunately, the three types of hardnesses—indentation, scratch and rebound—are not related to each other or inter-convertible. However, the different indentation hardness values measured are to some extent inter-convertible.

INDENTATION HARDNESS

When we pierce a block of wax and a block of wood with the tip of a nail, the nail will penetrate deeper into the wax than into wood. This is due to the higher resistance offered by wood. The factors which effect the depth of penetration of the nail can be seen to be :

    (a) the pressure used (load applied) for indentation,

    (b) the dimensions of the tip of the nail, and

    (c) the time for which the load is applied.

The depth of penetration can be an index of the hardness of the material if the above three factors are standardized. This is how indentation hardness is measured.

Brinell Hardness Test

Brinell hardness testing machine uses a hardened steel ball as the indenter. Balls of different diameter from 2.5 to 10 mm are employed with suitable loads. A schematic diagram of the Brinell hardness testing machine is given in Fig. 14.1 and a photograph of the same is shown in Fig. 14.2.

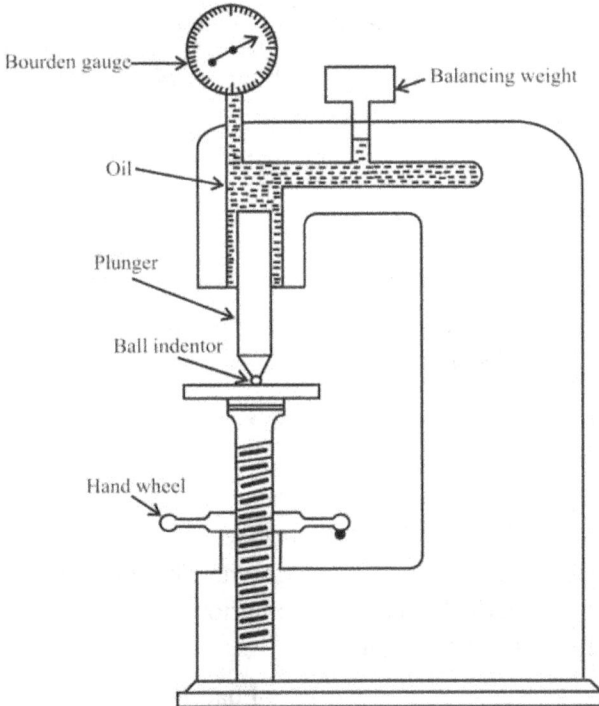

Fig. 14.1 Brinell hardness tester (hydraulic loading). Schematic diagram.

A hydraulic pump applies the load required for the specified time, i.e., 10 secs. The pump is operated by the tester himself in the older type of machines. Nowadays, electrically operated machines are available.

Fig. 14.2 Brinell hardness tester.
(*Courtesy* : Avery-Denison Ltd Leeds, U.K.)

After indenting, i.e., applying the load on the ball, for 10 secs, the load is removed and the indentation made is measured. In the Brinell machine, the surface area of the indentation is calculated and used as an index of the hardness of the metal. The surface area of the indentation is dependent upon the depth of penetration. The load applied (in Kgf) divided by the spherical area of indentation (in square millimeters) is taken as the Brinell Hardness Number.

$$\text{B.H. N.} = \frac{\text{Load } P}{\text{Spherical area of indentation}}.$$

Calculation of the spherical or surface area of indentation. The circle in Fig. 14.3 is the periphery of the indenter, whose diameter is D. The indented portion of the material shown shaded is the replica of a part of the surface of the indenter.

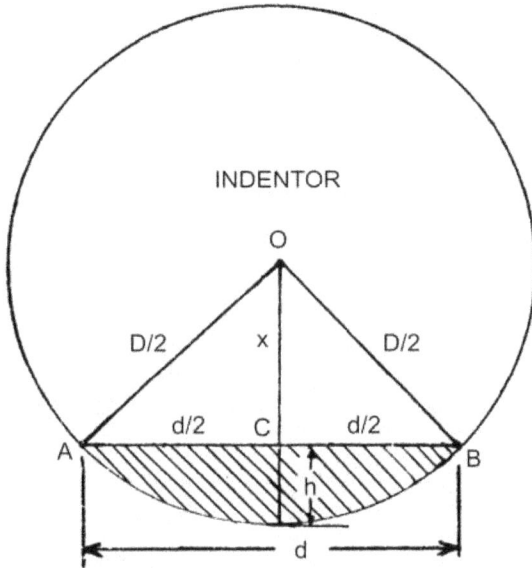

Fig. 14.3 Brinell Indentation

From mensuration principles, the surface area of a segment of a sphere is

$$= \pi D h \qquad \qquad ...(1)$$

where D = diameter of the sphere, and

h = height of the portion under indentation.

From Fig. 14.3, $h = \left\{ \dfrac{D}{2} - x \right\}.$ $\qquad \qquad ...(2)$

Again, from the right angled triangle OCB,

$$(D/2)^2 = X^2 + (d/2)^2.$$

or $x = \sqrt{\dfrac{D^2 - d^2}{4}}$

$$= \frac{1}{2} \sqrt{\left(D^2 - d^2 \right)} \qquad \qquad ...(3)$$

Substituting in (2) for 'x' we get,

$$h = \frac{D}{2} - \sqrt{\frac{\left(D^2 - d^2\right)}{2}}$$

$$= \frac{D - \sqrt{\left(D^2 - d^2\right)}}{2}.$$

Substituting in (1) for 'h', we get the spherical area of indentation

$$A = \frac{\pi D}{2}\left\{D - \sqrt{\left(D^2 - d^2\right)}\right\}.$$

The value of '*D*' is known and the value of '*d*' is measured with the help of a Brinell microscope (Figs. 14.4a and 14.4b). The hardness of the metal can either be calculated from the formula

$$\text{B.H.N.} = \frac{p}{\frac{\pi D}{2}\left(D - \sqrt{\left(D^2 - d^2\right)}\right)}$$

or by reading the standard charts which indicate the hardness number directly based upon the load applied and the diameter of the indentation obtained.

Fig. 14.4 a. The Brinell
Microscope.
(*Courtesy* : Avery-Denison, Ltd.,
Leeds, U.K.)

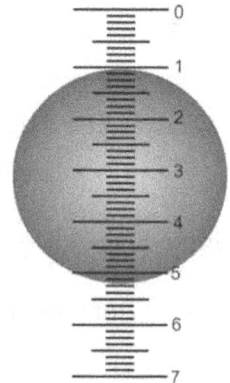

Fig 14.4 b. Brinell Indentation
under the Brinell Microscope.
(*Courtesy* : Avery-Denison Ltd.,
Leeds, U.K.)

Scope and Applicability

The Brinell hardness test makes use of a steel ball as an indenter. The hardness of the steel ball should be sufficiently higher than the material under test, otherwise the indenting ball itself will be deformed. Thus, the hardness of steels in the fully or surface hardened condition cannot be determined by this method. In other words, the Brinell Hardness Numbers are not reliable above 350. Further, depending upon the material under test, right selection of the ball and the load should be made.

From the relation

$$\text{B.H.N.} = \frac{P}{\frac{\pi D}{2}\left(D - \sqrt{\left(D^2 - d^2\right)}\right)}$$

We can see that there is the constant P/D^2,

If this is fixed, the indentations produced or the values of 'd' will be in a particular range. Hence, the ratio P/D^2 is maintained at different values in order to obtain measurable indentalions on the materials.

Usually P/D^2 = 30 for steels,

                          = 10 for non-ferrous metals,

                          = 2 for soft metals like lead.

The surface of the material should be even and smooth in order to obtain a clear indentation. Irregularities induce difficulties not only in the proper distribution of the load but also result in impressions with no well-defined contours. Thus, an accurate measurement of the diameter of the indentation becomes impossible.

Very thin specimens should not be tested by the Brinell method. When there is not sufficient material to back up, the region of indentation may become unduly large due to the least resistance offered to the ball. The thickness of the metal under test should be at least equal to the diameter of the indenter.

Similarly, the measurements at the edges and corners of the materials become difficult. Round objects should be filed out at a suitable place so as to obtain a plain surface and tested by placing them on U or V-shaped backing anvil.

The indenting ball should be frequently checked for its spherical shape and accuracy by finding out the hardness of the standard test block supplied along with the machine.

The Brinell hardness number is roughly related to the tensile strength. This empirical relationship is as follows:

$$\text{T.S.} = K \times \text{B.H.N. (tons/inch}^2)$$

where $K = 0.217$ for alloy steels and

0.22 for plain carbon steels.

For wrought light alloys, the relationship is as follows :

$$\text{T.S.} = 0.25 \times \text{B.H.N.} - 1 \text{ (tons/inch}^2)$$

The Brinell hardness number (B) is also related to the rebound hardness measured by Shore's scleroscope (S) as per Beeching's formula which is as follows :

$$S = 0.108\ B + 8$$

and $S = 0.1\ B + 15$ when S is more than 55.

[These (Beeching's formulae) are applicable for steel.]

Vickers Hardness Testing Machine

Vickers hardness machine also functions on the same principle as the Brinell hardness testing machine, but employs a square based pyramid made of diamond as the indenter. The size of the indenter is also very small. The included angle between the opposite faces of the pyramid is 136°.

This machine is more versatile than the Brinell hardness tester. Instead of changing the indenter as well as the loads depending upon the nature of the material tested, only the load is changed in the Vickers hardness tester. Varying loads from 1.0 Kgf. onwards up to 120 Kgf. are employed. Owing to the fineness and the small size of the indentation obtained, the specimen needs a glossy surface finish for testing. Generally the loading device is in the form of a pedal. The indenter and an objective to measure the indentation can be swung into the correct positions as required. A schematic diagram of the machine is given in Fig. 14.5.

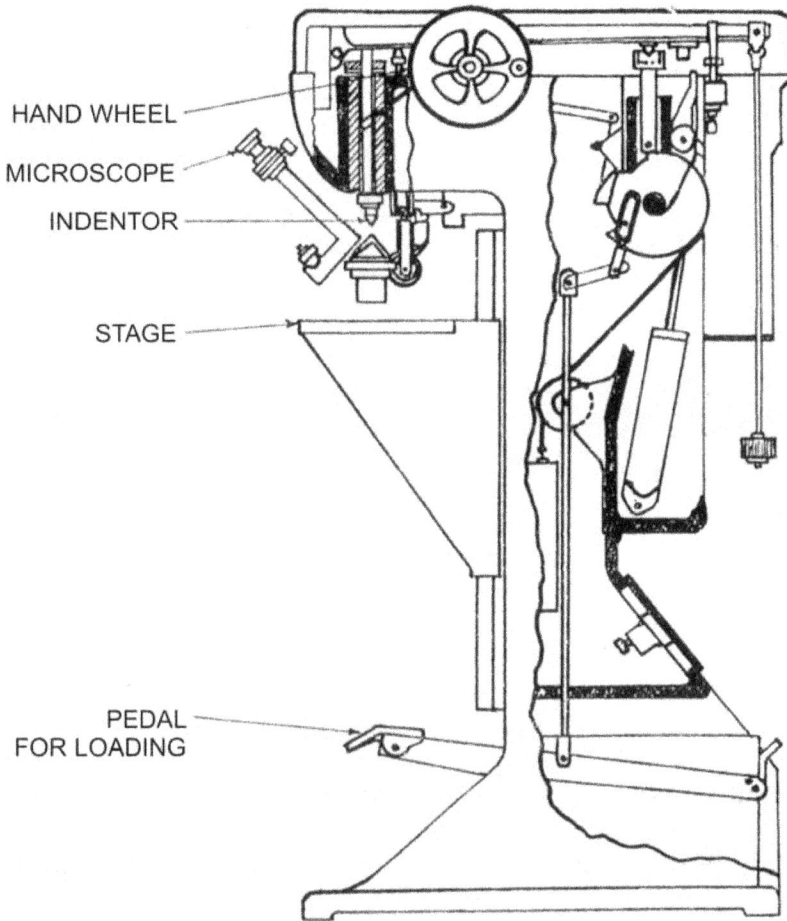

HAND WHEEL

MICROSCOPE

INDENTOR

STAGE

PEDAL
FOR LOADING

Fig. 14.5 Vickers hardness testing machine. (Schematic diagram.)

After indentation the size of the indentation is accurately measured by swivelling the microscope into position. The impression obtained will have a square sectional area (Fig. 14.6). The diagonal is measured using a crosswire focusing device in the optical equipment. In some types of hardness machines like the Galileo type, the indentation can be focused onto a graduated ground glass screen and measured.

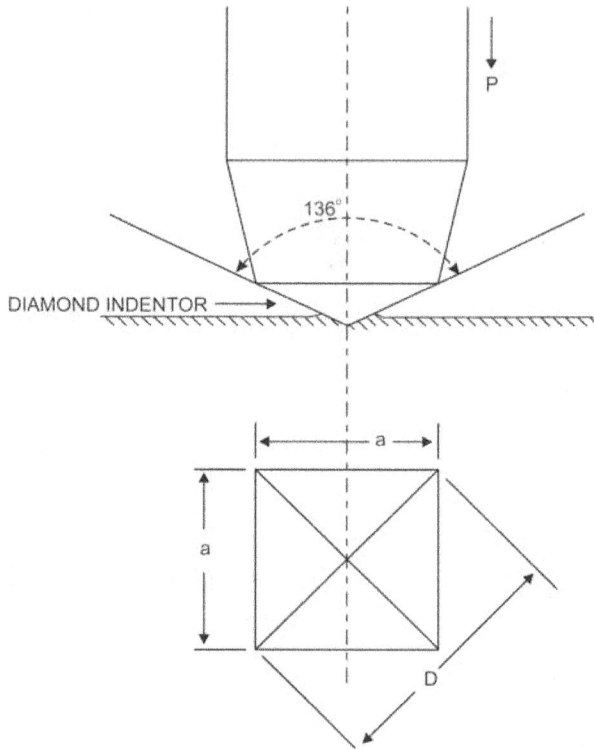

Fig. 14.6 Vickers indentation.

As in the case of Brinell hardness tester, the numerical value of Vickers hardness number is equal to the load applied, divided by the surface area of the pyramidal indentation. Thus the diamond pyramid hardness number

$$D.P.H. = P/A$$

where P = load applied, and

A = lateral area of pyramidal indentation in sq. mm.

The area of the pyramidal impression is calculated from the mensuration formula.

$$A = 4a\ S/2 \qquad\qquad ...(1)$$

where 'S' is the slant height of the impression.

From Fig. 14.6 it can be seen that

$$D^2 = 2a^2 \ \text{ or } \ a = D/\sqrt{2}$$

Thus, $A = (4D/\sqrt{2})(S/2)$

$\qquad = \sqrt{2}\,DS$ ...(2)

The included angle of the indentation is the same as the included angle of the indenter pyramid. This angle is $136°$ (Fig. 14.6). The slant height can then be expressed in terms of D as follows:

$$S = \frac{a}{2\sin 68}$$

$$= \frac{\dfrac{D}{\sqrt{2}}}{2\sin 68} \qquad\qquad ...(3)$$

Substituting the value of 'S' from (3) in (2) we get

$$A = \frac{D^2}{2\sin 68} \qquad\qquad ...(4)$$

If the value of 'A' as for (4) is substituted in the hardness number equation. Then,

$$\text{D.P.H.} = \frac{P}{D^2/2\sin 68}$$

$$= \frac{2.\sin 68.P}{D^2}$$

If the value of the constant 2 sin 68 is evaluated,

$$\text{D.P.H.} = 1.854\,P/D^2$$

where $P =$ applied load,

and $D =$ diagonal length of the impression.

Standard charts are supplied along with the Vickers machines, which give the Vickers hardness number depending upon the diagonal length of the impression at each load. Using these charts, the hardness number can be readily found.

The machine is the most accurate and versatile. Very high degree of polishing of the metal surface is required so that the impression can be perfectly focused and viewed and the diagonal accurately measured. Very careful handling of the indenter is recommended, besides periodic checking of the instrument against the standard steel blocks supplied alongwith.

Brinell cum Vickers Hardness Tester

Of late, machines capable of performing either Brinell or Vickers hardness tests are in the market. One such, Avery Denison (Type 6406), is illustrated in Fig. 14.7 a. It is supplied with hardened steel ball indentors of 1 and 2 mm diameter as well as the pyramidal diamond indentor. The indentation load can be varied by changing the weights at the back of the instrument. The specimen surface should be mirror-polished so that the indentation is accurately measured in the projected image on the ground glass screen.

Fig. 14.7 a. Model 6406 Vickers and Brinell hardness tester.
(*Courtesy* : Avery-Denison, Ltd., Leeds, U.K.)

The indenter as well as the image projection objective will come into position while loading and measuring, respectively, by means of the electric make-and-break mechanism inside the instrument. As usual, the machines are supplied with all the accessories like an anvil of larger diameter (to hold larger test pieces), slotted anvil (to support round stock), adjustable work rests (for lengthy objects), and calibrating (standard) Hardness blocks. The indentation is magnified and projected onto a ground glass screen fitted at the top of the machine. A vernier mechanism, provided in the screen, facilitates accurate measurement of the indentation (Fig. 14.7 b).

Fig. 14.7 b. Measuring screen showing magnified image of
Vickers impression.

The Brinell and Vickers hardness values are similar in their nature from the basic principles. The two values will tally up to the hardness number 400, and thereafter, the Vickers values are found to be more (see Fig. 14.8). However, in this range, i.e., above 400 Vickers hardness values are more reliable, as already mentioned above.

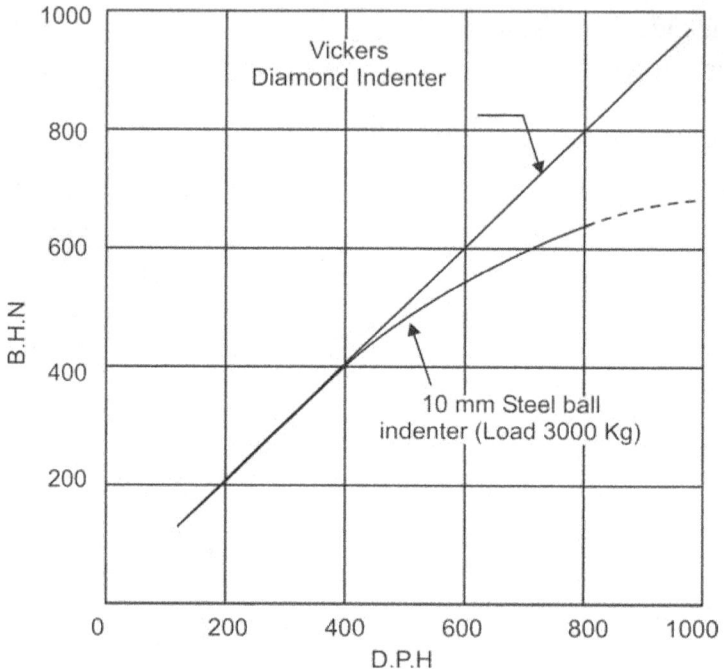

Fig. 14.8 Comparison of Brinell and Vickers hardness values.

Rockwell Hardness Tester

We have seen hitherto that when a metal is indented, it opposes the indentation. The degree of resistance offered by the metal is commensurate with its hardness. If the hardness is more, the resistance to indentation is more, and vice versa. Thus, we will have small indentations with hard metals and larger indentations with soft metals. Further, we have computed the surface area of indentation as an indirect index of the hardness.

Based on the same phenomenon, we can see that the depth of penetration is also inversely proportional to the material hardness. This can be utilized as a means of measuring the hardness. This concept was proposed in 1908 by Ludwig at Vienna.

Rockwell hardness tester is developed with the depth of penetration as the criterion for the hardness of the metal. A schematic diagram, underlining the principle of the working of the instrument, is given in Fig. 14.9.

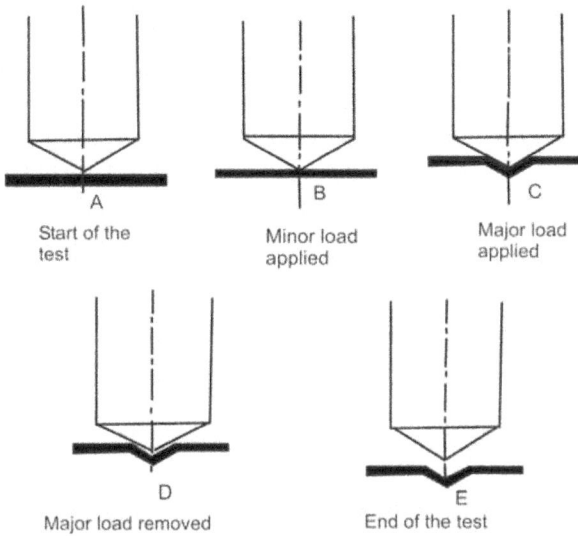

Fig. 14.9 The Rockwell indentation procedure.

Description of the machine. The photograph of a common type of Rockwell hardness tester is shown in Fig. 14.10. The machine consists of

Fig. 14.10 Special Rockwell for large parts.
(*Courtesy* : Wilson Instrument Division of A.C.C.O, Connecticut, U.S.A.)

an anvil which can be changed depending upon the shape of the specimen under test. Different anvils are illustrated in Fig. 14.11. The anvil can be moved up or down by turning the hand wheel which is situated at the bottom of the spindle. A loading lever is situated at the right hand side bottom portion of the machine. The loading may also be applied by simply operating a hand-lever which is just below the handwheel.

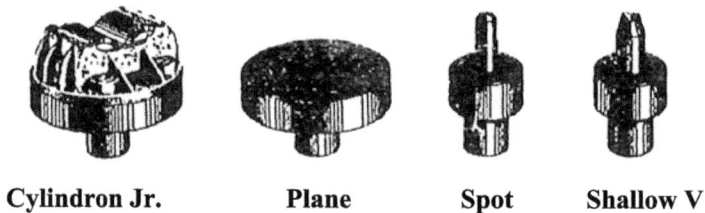

Cylindron Jr.   **Plane**   **Spot**   **Shallow V**

Fig. 14.11 Different types of anvils.
(*Courtesy* : Wilson Instrument Division of A.C.C.O., Connecticut, U.S.A)

The machine is a direct reading type. The dial consists of two pointers. The small pointer which is situated in the top left quarter of the dial proper indicates the application of the minor load of 10 Kgf. The anvil, with the specimen placed on it, is turned slowly so as to lift it upwards. The small pointer starts to move, once the specimen touches the indenter. The anvil is lifted up until the pointer comes to the stipulated position, viz., until it comes up to the black dot. This indicates that the minor load of 10 kgfs is acting upon the indenter.

The dial is then turned to the required position, i.e., the zero on the C-scale. The major load is applied by pressing the hand-lever (Fig. 14.9C).

Under the influence of the major load which is an increment over the already applied minor load, the indenter starts to go down into the specimen during the specified time. This can be seen from the dial. The pointer starts to move during the period of loading. After the pointer stops, the major as well as the minor loads are removed by turning the hand-lever at the right-hand side bottom in an anticlockwise direction. The reading on the relevant scale is the hardness of the metal (Fig. 14.9D). The action of the indenter under the influence of minor and major loads is schematically illustrated in Fig. 14.12.

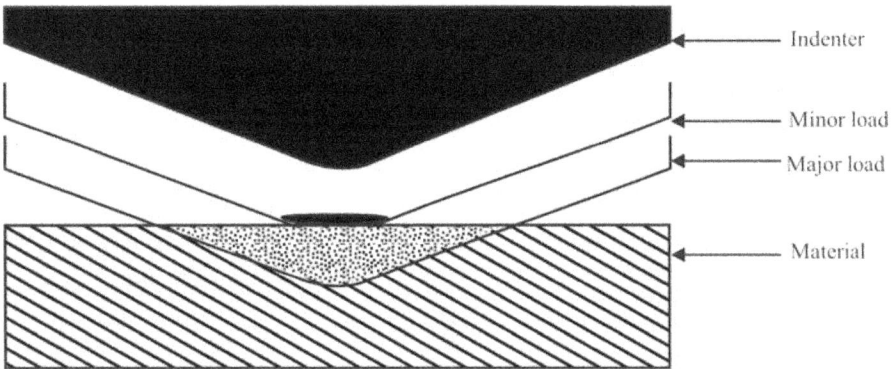

Fig. 14.12 Rockwell indentation.

The indenters and scales. The common indenters employed in the Rockwell hardness tester are a 1/16″ hardened steel ball and a diamond cone. When the former is employed, the major load is maintained at 100 Kgf and the hardness values read on the B-scale. The readings range from zero to 100. The B-scale is suitable to test plain carbon steels, and common non-ferrous metals and alloys like copper, brasses, bronzes, etc. However, for very soft metals like lead, indenter balls having 1/8″, 1/4″ and 1/2″ diameters are also supplied. When these balls are used, the major loads also will be altered and the hardness value is expressed on a different scale, e.g., instead of R_B, it is expressed as R_E when the 1/8″ diameter, ball indenter is used. It means that hardness value is designated by the E-scale. For testing very hard materials like hardened steels a diamond indenter is employed. It consists of a $120°$ conical-shaped diamond with a 0.2 mm diam. spherical indenting tip. This is also known by the patented name 'Brale'. With Brale, the major load consists of 150 Kgf. The hardness is read on the C-scale. A detailed account regarding the various hardness scales employed in the Rockwell hardness tester is given in a tabular form.

Rockwell Hardness Scales and Prefix Letters.*

Scale symbol and prefix letter	Indenter	Major load, Kg	Dial Numerals	Typical applications of scales
B*	Group one** 1/16-in. ball	100	Red	Copper alloys, Soft steels, aluminum alloys, Malleable iron.
C**	Diamond cone	150	Black	Steel, hard cast iron, pearlitic malleable iron, deep case-hardened steel
A	Group two Diamond cone	60	Black	Cemented carbides, thin shallow case-hardened steel
D	Diamond cone	100	Black	Thin steel, medium case-hardened steel
E	1/8-in. ball	100	Red	Cast iron, aluminum and magnesium alloys, bearing metals.
F	1/I6-in. ball	60	Red	Annealed copper alloys, thin soft sheet metals
G	1/I6-in. ball	150	Red	Phosphor bronze, beryllium copper, malleable iron.
H	1/8-in. ball	60	Red	Aluminum. Lead, zinc
K	1/8-in. ball	150	Red	
L	Group three 1/4-in, ball	60	Red	Bearing metal and other very soft or thin materials. Use smallest ball and heaviest load that does not give anvil effort.
M	1/4-in, ball	100	Red	
P	1/4-in, ball	150	Red	
R	1/2-in, ball	60	Red	
S	1/2-in, ball	100	Red	
V	1/2-in, ball	150	Red	

* Based on ASTM E 18.
** Commonly used scales and indenters.

Both the most commonly used Rockwell scales B and C consist of 0-100 divisions. But they are staggered in such a fashion that the B 30 is at C 0. It is made so, to avoid the negative hardness values on the B-scale if used to test very soft materials. This also facilitates in establishing that the highest hardness that can be measured with a 1/16″ diam. ball indenter is only B_{100} and for higher hardness the C-scale should be employed.

The actual usable range of the Brale is only from C_{20} onwards up to C_{70}. C_{20} corresponds to B_{97} and below that hardness level, it is not recommended to employ the Brale. In such cases, the shape irregularities at the base portion of the diamond cone begin to play an important role in giving false hardness values. Much care may not be exercised in shaping the bottom regions of the Brale as is exercised on the circular tip. On the other hand, if the C-scale is used for testing materials which are harder than C_{70}, the diamond cone may be damaged due to the high value of the load (150 Kgf) and the reaction of the extremely hard surface. In such instances the A-scale which employs a smaller load of 60 Kgf is better suited.

The useful range of the C-scale is practically from 40-70. This corresponds to the Vickers hardness number range of 410 to 1000. On the other hand when the A-scale is employed, the useful range is between 63-90, roughly corresponding to the Vickers hardness number range of 240 to 1400.

Advantages of the process

Listed below are the advantages of the process :

1. The process is relatively fast.
2. The hardness is directly read on the dial.
3. There is greater latitude for soft to hard materials.
4. A small indentation is left on the object.
5. Use of the initial minor load avoids the errors arising out of the uneven surface of the metal, and machine and table errors.

Disadvantages

The process has its drawbacks which are as follows :

1. The scale range is contracted.
2. The use of more than one scale to express the hardness.

3. Difficulty to readily convert the Rockwell hardness values either into Brinell or Vickers.

4. Errors are likely to be induced when the work is large and overhangs.

Rockwell Superficial Hardness Tester

This machine, shown in Fig. 14.13, is similar to the ordinary Rockwell machine. The indenters used are also the same, viz., the 1/16″ hardened steel ball and the conical based spherical diamond indenter. For softer materials the ball is used and for harder materials the Brale is used. When used in the superficial tester the Brale is designated as 'N Brale'.

Fig. 14.13 Rockwell superficial hardness tester.

(*Courtesy :* Wilson Instrument Division of A.C.C.O; Connecticut, USA)

The hardness reading dial too is similar to that of the ordinary machine. It consists of 100 equal divisions. Each division represents 0.001 mm vertical motion of the penetrator. The dial is very sensitive to the motion of the indenter as the impressions obtained are very small.

Rockwell Superficial-Hardness Scales*

Major load, kg	Scale symbols				
	N scale, Diamond cone	T scale, 1/16-in. ball	W scale, 1/8-in. ball	X scale, 1/4-in. ball	Y scale, 1/2-in. ball
15	15 N	15 T	15 W	15 X	15 Y
30	30 N	30 T	30 W	30 X	30 Y
45	45 N	45 T	45 W	45 X	45 Y

*Based on ASTM E 18.

A minor load of 3 Kgf is used and the major load may be 15, 30 or 45 Kgf as against the 10 Kgf and 60, 100 and 150 Kgf, respectively, in the ordinary machine. Because of this, superficial hardness scales are devised. These are defined by the indenter used. Thus when the N Brale is used the hardness is read as a N-number. Besides, the major load employed is also mentioned. Thus, if the N Brale is used with a 30 Kgf. major load the scale is prefixed as 30N and so on.

Standardization

The machine is frequently checked by finding the hardness of the standard plates supplied along with it.

Application

The machine is particularly suitable when the hardness of the metallic region to be determined is shallow such as case-hardened cases, electroplatings, etc.

Monotron Hardness Tester

In this machine a standard 0.75 mm hemispherical diamond indenter is employed. It is forced into the metal upto a standard depth of 0.05 mm. The load required to force the penetrator is dependent upon the hardness of the material.

The machine consists of two circular scales. The top dial indicates the load. The hardness number is read directly on the top scale with the indenter still in the specimen. The scale is graduated in Kgf. and Brinnell units. For testing softer materiats, large indenters made of tungsten carbide, in diameters of 1.53 or 2.5 mm. are available. When these are employed, the hardnesses are referred to as of the M_3 and M_4 scales respectively.

The bottom dial is for reading the depth of penetration. There are two scales on the dial, each being divided into 100 equal parts. One scale, which is marked in the clockwise fashion, is for routine testing, i.e., the constant depth indenting. The other scale, marked in the anticlockwise direction, is for performing specific tests, such as measuring the depth of indentation or flow of the metal at a constant load of indentation.

The machine can also be used to test the hardnesses of thin layers like the carburised or nitrided cases and thin strip. In such cases where a deeper impression is undesirable, the penetration may be done up to a fraction of the specified 0.0018″, and the noted hardness value multiplied by the numerical in the denominator of the fraction. If an indentation of 0.0009″ depth is made and the load noted, the hardness is evaluated as the load multiplied by 2.

Instead of the Knoop indenter, an ordinary Vickers type pyramid may also be employed for indentation. In such a case the hardness calculation is done as in the Vickers hardness machine.

Precautions

The instrument is very sensitive and every care should be taken to see that the measurment of the diagonal is very accurately done. Since only the projected area of the indentation is taken into account, the measurement should be made exactly at right angles to the surface under test. To achieve this, the bottom portion of the specimen too should be very carefully prepared so that it rests properly on the stage of the instrument.

Mayer's Hardness

It is more rational to consider the projected area of indentation than the actual surface area in the calculation of the hardness. This idea was put forth by Meyer in 1908. According to him, the following empirical relationship exists between the applied load and the size of the indentation,

$$P = Kd^{n^1}$$

where $P =$ load applied in Kgs,

$K =$ a material constant based upon the resistance of the metal,

$n^1 =$ a material constant based upon the strain hardening of the metal,

$d =$ diameter of the indentation in mm.

If log p is plotted against log d, a straight line graph is obtained. The slope of the straight line is n^1 and K is the value of P when $d = 1$. The value of n^1 for fully annealed metals is found to be about 2.5 and for fully strain-hardened metals, about 2.0.

It can be proved that the value of P/d^2 is a constant in this case also, i.e., when the projected area of indentation is considered instead of the surface area for hardness purposes. The manufacture of this type of hardness testers is not in vogue now.

Tukon Hardness Tester

This is a special purpose machine, used for determining the hardness of extremely thin materials. Nitrided or cyanided cases and decarburised surfaces of steel and the microconstituents in an alloy can be tested by this instrument. Commonly, the depth of penetration of the indentation does not exceed one micron in this test.

The instrument shown in Fig. 14.14 is a complicated yet precision built apparatus. Except the measurement of the indentation, all the operations

Fig. 14.14 Tukon hardness tester.
(*Courtesy* : Wilson Instrument Division of A.C.C.O., Connecticut, U.S.A.)

are performed automatically by electronic control devices. To test the hardness of a microconstituent in an alloy, the polished and etched specimen is placed on the stage of the instrument. The phase under test whose hardness is being determined, is accurately focused and the specimen is lifted up by moving the stage upwards just below the indenter. Then the instrument is put on by pressing the button. The automatic mechanism functions and performs the following operations:

   (i)   further movement of the specimen to touch the indenter,

   (ii) stopping the upward movement and applying the load for the required time,

   (iii) upward movement of the indenter after loading, and

   (iv) lowering of the stage and thereby the specimen to the initial position.

The Knoop indenter, which is generally used, is a typically shaped diamond pyramid. The pyramid is cut to an included longitudinal angle of 172° $30'$ and transverse angle of 130° $0'$, within a 1% tolerance. This facilitates in obtaining an impression which is rhombic in shape with the longer diagonal approximately seven times the shorter diagonal. The indenter and the impressions obtained are shown in Fig. 14.15.

Hardness Test	Brinell	Rock well		Vickers	Knoop	Monotron
Material of the indenter	Steel	Diamond	Steel	Diamond	Diamond	Diamond
Indenter shape	Sphere	Cone	Sphere	Square pyramid	Rhombo pyramid	Hemi sphere
Dimensions of the indenter	$D = 10$	$\theta = 120^\circ$	$D = 1/16''$ $1/18''$ $1/4''$ or $1/16''$	$\theta = 136^\circ$	$\alpha = 130^\circ$ $\theta = 172^\circ 30'$ $172^\circ\ 30'$	$D = 0.75$

Fig. 14.15 a. Specifications for hardness indenters.

(From A.S.M. Metals Handbook, 1948.)

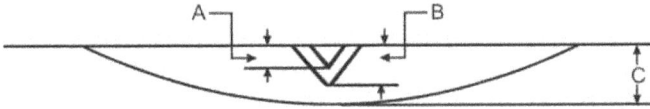

A. Superficial rockwell N diamond cone 30 Kgf. Load 0.0018″

B. Rockwell C scale brale 150 Kgf. Load 0.0052″

C. Brinell 10mm. Ball 3000 Kgf. Load 0.01″

Fig. 14.15 b. Comparative impressions in steel (R_e 39) using Brinell, Rockwell and Rockwell Superficial Testers. (Magnified approximately 30 times.) (From Lysaught, V.E., Indentation Hardness Testing, Reinhold, N.Y., 1949.)

The longitudinal diagonal is the one measured in the impression. The numerical value of the Knoop hardness number is given by the formula:

$$\text{Knoop Hardness Number} = P/A_p$$

where P = applied load in Kgf.

A_p = projected area of the indentation.

The projected area of the impression in the case of an indenter with perfect included angles is equal to

$$wl/2$$

where w = length of the shorter diagonal,

and l = length of the larger diagonal.

From the consideration of the included angles,

$$\frac{\tan 65°}{w} = \frac{\tan 86° \ 15'}{l}$$

Substituting the numerical values,

$$l = 7.114.w.$$

or $w = 0.14056 \, l$

Thus $A_p = \dfrac{0.14056 \, l^2}{2}$

$$= 0.07028 \, l^2$$

The value, 0.07028, is known as the Knoop indenter constant C_P. This denotes that if the indenter in use is having slight dimensional deviations, the actual value for C_P should be used rather than the theoretical value of 0.07028. The value of C_P is supplied by the manufacturers.

Thus
$$KHN = \frac{P}{C_p L^2}$$

where   KHN = Knoop hardness number,

P = load in Kgf,

L = the longitudinal diagonal of the impression, and

C_P = Constant supplied by the manufacturers of the machine.

Instead of using the Knoop indentor, an ordinary Vickers type pyramid indentor may also be employed in the machine. In such a case the hardness calculation should be performed as in the case of the Vickers hardness testing.

Precautions

The instrument is very sensitive and great care should be taken to see that the measurement of the diagonal is made accurately. Since only the projected area of the indentation is taken into account, the measurement should be exactly at right angles to the surface under test. To achieve this, the bottom portion of the specimen should also be very carefully prepared, so that it rests properly on the stage of the instrument.

Rockwell Microficial Hardness Tester

The principle of the Rockwell microficial hardness tester is based on the well-known Rockwell test. The loads applied are much lighter and the depth measurement system is more sensitive in view of the shallow indentations (as shallow as $2\mu m$). This shallow indentation is the result of the microficial indentor which is basically a Vickers indenter with an accurately ground flat tip. This flat tip limits the total depth of penetration without a loss in sensitivity. The principle of working is illustrated in Fig. 14.16.

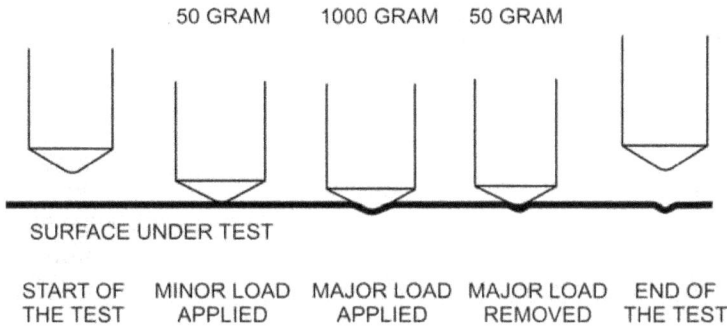

Fig. 14.16 Rockwell microficial hardness test.

The microficial tester is shown in Fig. 14.17. The equipment is the latest addition to the microhardness instruments and is of a direct reading type. Further, the whole cycle of testing takes just 15 seconds.

Fig. 14.17 Rockwell microficial hardness tester.
(*Courtesy* : Wilson Instrument Division of A.C.C.O., Connecticut, U.S.A)

Accuracy of the Rockwell Hardness Tester

A detailed survey was undertaken by a sub-committee of the British Standards Institution in the early sixties. It was reported that every three out of four Rockwell hardness testers working in the industry were giving non-repeatable values of hardness. Later on, the National Physical Laboratory (U.K.) had undertaken detailed investigation and came to the conclusion that the reasons for the lack of repeatability of hardness values of the industrial machines were poor maintenance of the machines and improper testing methods like not supporting the work under test properly. It was concluded that the principle of Rockwell testing was inherently a very accurate method.

It is prescribed by the British Standards Institution that in periodic checking with standard blocks, the machines should give hardness values in the range of ± 2 readings for the B-scale and ± 1.5 readings for the C-scale.

The Indian Standard Specification I.S. 3804 : 1966 stipulates that the machine should give readings with standard blocks within 3% of the actual hardness (of the block) for the C-scale and 6% for the B-scale.

Scratch hardness

This is by far the oldest hardness testing method. Though seldom used except for mineralogical purposes, the testing is very simple according to a hardness scale known after Mohs. It consists of ten minerals arranged in increasing order of hardness. The principle behind this method of testing is that when two materials are scratched against each other, the harder of the two scratches the other. The material under test is scratched with the mineral number 1. If the mineral fails to scratch the material, then the next harder mineral, i.e., mineral number 2, is used, and so on until the material is scratched by a mineral. Suppose the mineral which scratches is the one numbered 5. Then the hardness of the material is expressed as lower than 5 and higher than 4. The Mohs scale of hardness is as follows:

1. Talc	2. Gypsum
3. Calcite	4. Fluorite
5. Apatite	6. Orthoclase (or feldspar)
7. Quartz	8. Topaz
9. Corundum	10. Diamond.

Small pieces, of about two cm. size, are arranged in numbered chambers in a wooden box resembling the weight box of a conventional balance. This forms an important appliance for testing in geological laboratories, besides serving as a portable unit for geological expeditions.

File Test

This is a very useful method of testing the abrasion hardness of metals. A workshop file is all that is used. The material under test is just filed and it is noted whether the file cuts it or not.

On the Mohs scale, the hardness of a file can be put at 6.

An important point to be noted here is that, undue pressures of filing should be avoided, or else the material as well as the file may be cut off.

Penknife Test

An ordinary penknife which is commonly placed in the key purse is used to scratch the material in this test. The hardness of the penknife can be put in the range of 4 to 6 on the Mohs scale.

The Microcharacter Hardness Test

The instrument appears like an ordinary metallurgical bench microscope. In fact, it is so in that the whole thing is an attachment made to a metallurgical microscope.

To test the scratch hardness of a microconstituent, the polished and etched specimen is placed on the stage of the instrument. The exact phase or microconstituent to be scratched is focused and the microcharacter attachment is fixed to the objective. The whole attachment is simple and consists of a well-balanced lever to the bottom of which a sharp diamond-cutting tool is fixed. The lever arm is properly weighted and lowered on to the surface under test. When the micrometer arrangement of the microscope stage is moved, the specimen rigidly fixed on to it also moves against the diamond pointer. Consequently a scratch is made on its surface. The width of the scratch (made on the surface) is governed by the hardness of the material. The width of the scratch differs from region to region, it being wider on the regions which are soft and narrow on the regions which are hard.

Watch oil is used to coat the diamond before cutting so that the cutting action will be smooth. The specimen is removed, cleaned with an organic solvent (cleaning reagent like xylol or carbon tetrachloride) and the width of the scratch on the required phase is measured.

The microhardness 'K' of the phase is obtained by the following empirical formula:

$$K = \frac{10,000}{\lambda^2}$$

where λ is the width of the scratch in microns.

The figure 10,000 is just a conventional one, to avoid fractional values of 'K'.

Generally the load applied to the diamond is 3 gmf. The diamond cutter is attached to a spring sort of a device so that it will scratch the surface effectively, especially when relatively harder and softer regions are present side by side.

REBOUND HARDNFSS

When a standard hammer weight with a diamond tip is made to fall on to the surface of a metal, there will be a great instantaneous load at the point of impact. The kinetic energy of impact is expended in making an indentation at the spot and in rebound of the hammer. If the metal is hard, the indentation made will be small for it yields less, and the height to which the hammer rebounds will be more, and vice versa. If the height of fall of the hammer and its weight are maintained as constant, the height of the rebound of the hammer will give us an idea of the hardness of the metal. Shore's scleroscope operates on this principle.

The instrument essentially consists of a graduated glass tube through which a 2.6 gm. standard hammer falls through a standard height. The whole length of the glass tube is graduated into 100 or 140 equal parts. The height of the rebound may be noted thus:

(i) By watching the top of the hammer itself.

(ii) By means of a magnifying glass and pointer attached to the pipe which are moved to the expected region of rebound. This is done by preliminary trials.

(iii) By noting the number on a circular dial at the top of the instrument.

In all the instruments, the glass tube is aligned to be perfectly vertical. The two important models of Shore's scleroscope, the C_2 and the D are illustrated in Figs. 14.18 and 14.19. In the model D, the circular scale indicates the scleroscope number, in addition to giving conversion into other types of hardness scales.

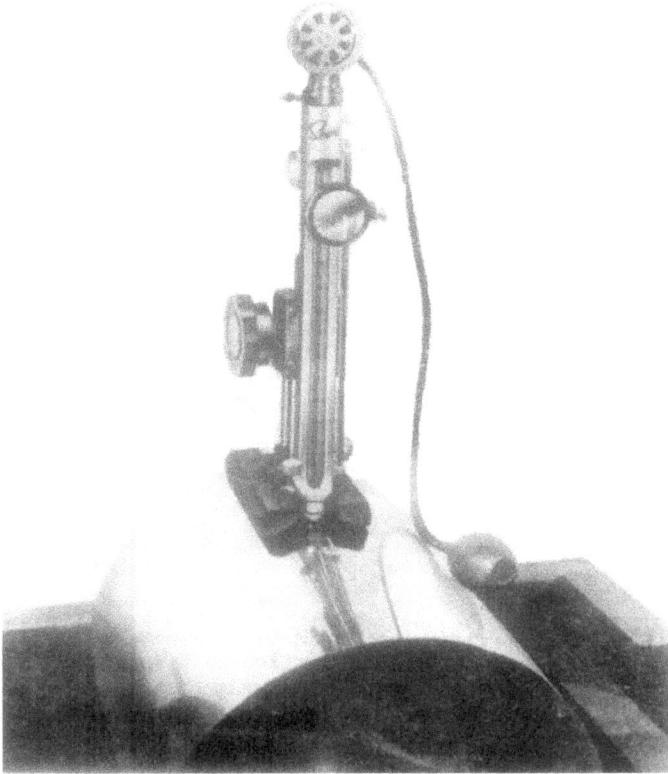

Fig. 14.18 Shore's scleroscope (Model C$_2$).
(*Courtesy* : The Shore Instrument & Mfg. Company Inc., Jamaica, N.Y., U.S.A.)

Sources of Error

1. The surface condition of the metal under test. If the surface is not perfectly plain, a smaller amount of rebound will occur. Generally it is necessary that the surface of the metal should be more perfect than is necessary in the case of either the Brinell or Rockwell test.

2. The specimen should be perfectly clamped to the base of the instrument, otherwise the rebound will not be proper due to energy losses on account of the movement of the specimen itself. This can be judged by the sound of the impact of the hammer while testing. The sound should be the same as a characteristic hammer blow on metal.

3. Decarburised layers, dust, grease, etc., should not be present on the surface.

Fig. 14.19 Shore's scleroscope (Model D).
(*Courtesy* : The Shore Instrument & Mfg. Company Inc., Jamaica, N.Y., U.S.A.)

Advantages

  (i)   The procedure is very short and simple.

 (ii)   The impression left on the object is very small.

(iii)   The instrument is portable.

Application

For routine inspection work, especially when 100% testing is necessary, the instrument is invaluable. Very thin sections can also be tested.

Limitations

In this instrument the rebound property of the hard surface is made use of. Since rebound is a characteristic of elastic materials also, elastic materials like rubber, etc., should not be tested by this instrument and the values obtained thereon should not be compared to the hardness of metals. The very nature of the material under test should be similar. Then only can a comparison of the values obtained on testing, be made.

Herberts Pendulum Hardness

This instrument not only determines the hardness of a metal manifested as the resistance to indentation but also as the resistance to work hardening.

In brief, the machine consists of a swinging weight in the shape of bracket. The weight of this pendulum is 4 Kgf. The centre of gravity of the pendulum is designed to act at its top bottom. The pendulum has a length of approximately 30 cm and is capable of accommodating articles of 1,500 mm width or 2,000 mm diameter. A one mm diameter hardened steel ball is fixed in a suitable chuck beneath the pendulum. A skeleton sketch of the instrument is illustrated in Fig. 14.20.

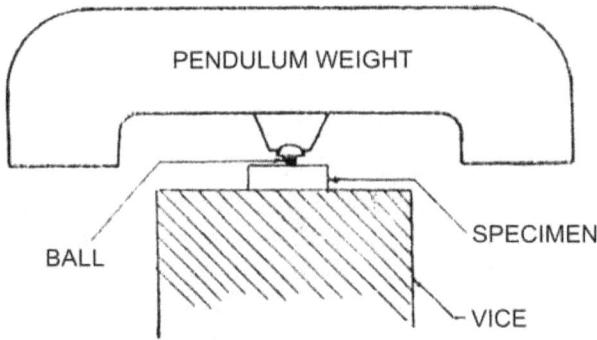

Fig. 14.20 Herbert's pendulum hardness tester.

Time Test

This determines the hardness of the metal. The time required for 5 oscillations of the pendulum is noted and the period is calculated as in the case of the simple pendulum. This is called the time hardness number. There is a factor by which the period can be multiplied to obtain the Brinell hardness number.

The time period will be short if the hardness of the metal is more. When the metal is hard, the indenter ball does not penetrate or get embedded into the metal. Consequently, there will be less resistance to the swinging of the pendulum. On the other hand, when the metal surface is soft, higher values of time period are recorded.

Scale test

The pendulum is tilted through a particular (specified) angle and released and the extent to which it swings on the other side is measured. The angle thus recorded is referred to as the scale hardness number. As explained above the angle of swing depends upon the energy absorbed at the point of contact. Thus, it is a measure of the resistance of the metal to plastic deformation.

From the above two, the 'flow hardness number' is derived, by dividing the scale hardness number by the time hardness number.

The Herbert's Pendulum Hardness (Number) tester is a special purpose machine. Its application is, however, mostly confined to the research laboratory.

A dynamic hardness tester which is very practically usable, even during the processing of a material is the Poldi Hardness tester. The instrument is handy and can be taken to the job when it is unwieldy and cannot be carried to the conventional hardness tester and mounted on its table for testing. It is a simple implement consisting of a hardened steel ball, located firmly in a tubular holder. A movable plunger rigidly held by means of a spring works in the tubular holder. A gap is provided between the ball and the plunger's bottom and into which a 'Standard' rectangular bar is inserted. This bar is hardened to a predetermined value. It has a tapered end to facilitate easy insertion into the space. The standard bar is held in the proper position due to the action of the spring provided in the tubular ball holder (Fig. 14.21).

The surface of the material under test is polished enough to enable the tester to accurately measure the indentation. The process is very simple. The indenter (i.e., the ball holder) is held vertically over the material surface (under test) with the standard bar inserted into it, the ball touching the surface. A hammer blow is given to the plunger from the top. The blow leads to the indentations on the test piece as well as on the standard bar. Both the indentations are measured accurately using a microscope, similar to the one used in Brinell hardness test. The degree of accuracy required is of the order of 0.1 mm.

STANDARD BAR

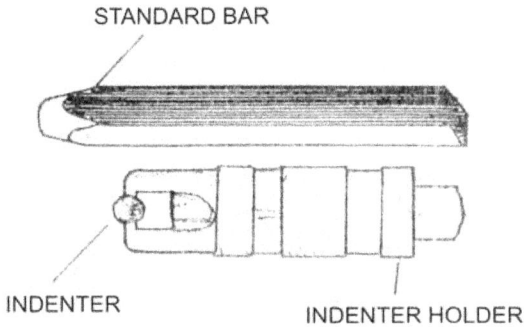

INDENTER

INDENTER HOLDER

Fig. 14.21

Tensile strength and Brinell hardness number of the material tested can be found out using the standard tables supplied alongwith the instrument. Separate tables are available for steels–annealed state and hardened or hardened and tempered state. Thus, to determine the hardness of the steel, it is necessary that its condition is known. However, even if it is not known, first the tables provided for natural or annealed state are to be used for steels having hardness upto 360. If the hardness exceeds 360, the other tables are to be used.

Separate tables are provided for other metals like copper, aluminium, and alloys like brass, bronze etc. However, these tables do not show the tensile strengths for, the relation between the Brinell hardness number and the tensile strength for these materials is precisely not established. For brass also, two sets of tables are provided, one for castings and the other for rolled brass.

Procedure

The tables consist of vertical and horizontal columns, giving the diameters of indentations on the specimen and those on the standard test bar respectively. Where the horizontal and vertical columns intersect, the corresponding square (in the table) gives the Brinell hardness number and tensile strength of the material.

If a correction factor is given on the standard test bar, the value read from the table is multiplied by the correction factor.

Precautions

1. The hammer used is an ordinary hand hammer of 1/2 Kgf.
2. The blow given should be normal, not very severe nor very mild.
3. The ball holder should be held vertically on the specimen.

4. If the indentation obtained is not circular, the average diameter should be taken to compute the hardness.

5. The standard test bar should be so inserted that the ball rests at a minimum distance of 15 mm for any previous indentation.

6. It should be noted that the Poldi hardness is no substitute for the Bench hardness tester and the values obtained are within 10% of the accurately measured hardness of the material.

HOT HARDNESS

The hardness of metal, just as it is related to the strength at the room temperature, is also found to be related to the high temperature strength at the concerned temperatures (Fig. 14.22). With the advent of the missile era and the prominence given to high temperature materials, the prediction of creep strength may be possible to a certain extent with the knowledge of the high temperature hardness. Methods of high temperature hardness testing are evolved by using a Vickers machine with a sapphire indenter, the indentation being performed in vacuum. There occurs a linear relation between the hot hardness and the strength of the metal; both varying (decreasing) almost parallelly with increasing temperature. High temperature Rockwell hardness testers are also in the market (Fig. 14.23).

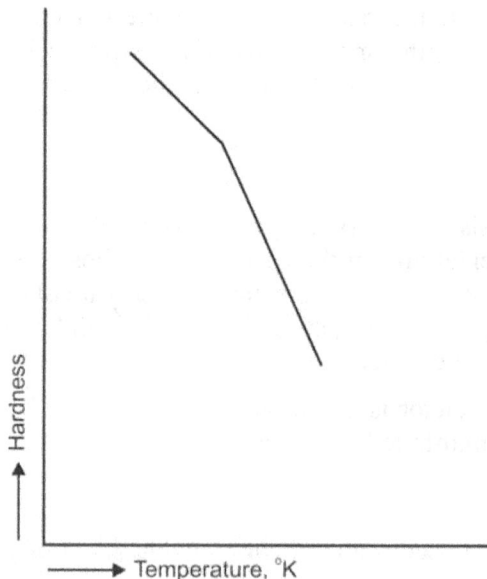

Fig. 14.22 Relation between hardness and high-temperature strength.

Fig. 14.23 High-temperature Rockwell hardness tester.
(*Courtesy* : Wilson Instrument Division of A.C.C.O., Connecticut, U.S.A.)

References

Westerbrook, T. H., *Trans ASM.*, Yol. 45, pp. 221-248, 1953.

Simons. Eric N., *Mechanical Testing of Metallic Materials*, Pitman.

IS: 1500–1968, Method for Brinell hardness test for steel.

IS: 1789–1961, Method for Brinell hardness test for grey cast iron.

IS: 1790–1961, Method for Brinell hardness test for light metals and their alloys.

IS: 3054–1965, Method for Brinell hardness test for copper and copper alloys.

IS: 1586–1968, Method for Rockwell hardness test (Band *c* scales) for steel.

IS: 1501–1968, Vickers hardness test for steel.

IS: 1810–1960, Vickers hardness test for light metals and their alloys.

IS: 2866–1965, Vickers hardness test for copper and copper alloys.

IS: 3754–1967, Calibration of standard blocks to be used for Rockwell B and C scale hardness testing machines.

IS: 4258–1967, Hardness conversion tables for metals.

Index

www.ingramcontent.com/pod-product-compliance
Lightning Source LLC
Chambersburg PA
CBHW050524190326
41458CB00005B/1651